Trial of an Iconic Species

Other Books by Scott Renyard

Illustrated Screenplays

Who Killed Miracle? (2022)

The Pristine Coast (2023)

The Unofficial Trial of Alexandra Morton (2023)

The Herring People (2023)

Screenplay Collections

Pressure Point: A series of mishandled events (forthcoming 2024)

Children's Books

The Flag That Flew Up (2021)

Trial
of an
Iconic Species

an illustrated
screenplay

Scott Renyard

juggernaut CLASSICS

Published by Juggernaut Classics Inc.

Contact: scott@juggernautpictures.ca

ISBN: 978-1-998836-56-7 (softcover)
ISBN: 978-1-998836-57-4 (eBook)

Cover: Original painting by Marcus Bowcott
Photo on page 2 by Anissa Reed

Edited by Lesley Cameron

Cover design by Rob Neilson and Jan Westendorp
Book design by Jan Westendorp/katodesignandphoto.ca

Juggernaut Classics Inc.

Contents

Introduction

The evidentiary hearings for the Cohen Inquiry, a multimillion dollar Canadian inquiry into the decline of Fraser River sockeye salmon populations, began on October 25, 2010. My first day of filming, however, was December 13, 2010. By that point there had been 14 days of testimony. If I had known earlier that there was a camera position available, I would have been there right from the start. As soon as I found out it was an option, though, I obtained permission to create a visual record of the proceedings with my camera (see *The Unofficial Trial of Alexandra Morton*) and was very excited to be contributing in my own way to the investigation into why many Fraser River sockeye populations were going extinct. As I listened to the different stakeholders and the cutting-edge science that was discussed, I soon felt like I was recording a multi-layered story with all the twists and turns you would find in a classic Hollywood courtroom drama.

I spent the better part of nine years at university studying for science degrees, but I never considered science to be as politically charged as it seemed to be during the Cohen Inquiry. Sure, there were always debates in science circles, but I suppose back then I hadn't yet been exposed to the politics that surround natural resources outside the ivory towers of university life. The goal of science, to my mind at least, was to add to the knowledge base and find solutions to problems. But it was clear from my first few days of recording the evidentiary hearings that scientific investigation was a lot messier than I could ever have imagined. Scientific inquiry involving fish, and sockeye salmon in particular, appeared to be an exceptionally hot political potato. During this time, scientists whose work in this field was perceived by the Conservative government to jeopardize economic interests faced restrictions on reporting their findings or risked losing their jobs—and sometimes both.

The Cohen Inquiry had a broad mandate and investigated a wide variety of topics, including, but not limited to, basic ecology, predation, impacts on habitats and the Fraser River watershed, fishing, disease, climate change, salmon policy, and fisheries management. I realized that many of the issues being presented and discussed had to do with old conflicts and were often tied to fishing rights and allocations. These issues are important, and catching too many fish has a huge impact on the sustainability of sockeye salmon populations, but I was particularly interested in the specific reasons for the sockeye salmon declines that began in earnest in the 1990s. I had my own suspicions about the reasons, which I reported in my documentary *The Pristine Coast* in 2014. I still couldn't get my head around how large government ministries filled with scientists and very educated managers appeared to be unable—or unwilling—to dig deeper to find a solution—or at least, to do so publicly. So I decided to make another film, one focused on the politics of science, which seemed to be the primary reason the declines in sockeye populations were allowed to go on for so long without someone taking action.

By the end of the Inquiry I had recorded 23,641 minutes of testimony. It was a daunting task to sit down and work on the material and turn it into a film. Slowly, over the months, I started to see that some key questions were being asked, but the responses were not always clear. Panellists often stumbled over their words and gave answers that were general and even vague. I often found myself wondering if the speaker actually knew the answer—and if not, why not? The body language of some panellists made me think that they were being very cautious about the answers they offered up. Some were defensive. It felt like some were afraid that they could lose their job if they said the wrong thing. And they were probably right.

Many of Gregory McDade's questions probed why the Department of Fisheries and Oceans (DFO) wasn't clear on the causes of wild salmon declines. After several decades and hundreds of millions of tax dollars, why don't we know the answers to some basic questions? As more testimony came out, it became obvious to me that directives from politicians over the last few decades have played an increasingly key role in the functioning of the DFO. In fact, a former DFO employee stated as much in my film *The Pristine Coast* when he said that since Brian Mulroney's time in office, the DFO had become very politicized. And it seemed to me that little had changed in that respect. If anything, things were likely worse than they were before. If the government perceives that tax revenue could be lost, the DFO, instead of solving or answering difficult science problems via the top-level research it has long been known for, resorts to issue management. Some say that the DFO has been drastically underfunded for a very long time and that the underfunding has, in part, caused the department to fail to deliver on its mandate. Or perhaps it has squandered the limited resources it did have on promoting the wrong things by becoming a cheerleader for aquaculture. In any case, the DFO was no longer a fisheries department but in many ways had become a public relations department. So, from this perspective, the Cohen Inquiry did what it was meant to do. It exposed the government's short-term focus on economic interests and failure to do the right long-term thing for the fish. As a result, there were some quite dramatic and revealing moments. For example, when Dr. Scott Hinch revealed why the word "virus" wasn't mentioned in his report to the Inquiry and why he could now talk about it, it was a pinnacle moment. I could even imagine the movie score rising to the climax moment as I heard him say, "I nonetheless inserted the reference in so that it would get into the document so that we could talk about it."

What the Hinch testimony (along with other testimonies) also revealed was that there were a lot of rules about what information was allowed to be evidence for the Inquiry. This meant that key findings that could have had a substantial effect on the Commissioner's recommendations nearly did not make it into the Inquiry. Who would've thought that such drama would find its way into a Canadian inquiry about fish?

The main points of interest that emerged from the science testimony didn't really fit the usual three-act structure for a film. But they did create a narrative in which the stakes seemed to get higher and higher as we moved closer to what could be the answer to the main question that triggered the Cohen Inquiry in the first place. After I had spent several hours mulling over the material, the story began to feel more like a book than a film. That's when I decided to break the film, which I called *Trial of an Iconic Species*, into 12 chapters, with each one asking a key question. Usually a book forms the basis for a movie; in this case it was the other way around.

With hindsight, I suppose I assumed that the main reason for the often cloak-and-dagger-like testimony was that the damage that was being done to wild salmon was a huge liability for the aquaculture industry. I also assumed that the government was protecting what it saw as a source of tax revenue. Avoiding any kind of a shutdown to protect wild sockeye salmon therefore seemed to me to be the government's main motivation in siding with the aquaculture industry. But even a quick back-of-the-envelope calculation revealed that it was a bad deal for Canada to turn its back on one of its most valuable natural resources instead of protecting it. Why would our governments want to sacrifice 60,000 wild salmon–related jobs for fewer than 2,000 aquaculture jobs? I'm not an economist, but doesn't wider participation in a resource make for a better, more resilient economy? It was a real head-scratcher.

It was chapter 11 of the film before I had my ah-ha moment. Dr. Gary Marty revealed in his testimony that one of his clients is the Canadian

Food Inspection Agency (CFIA) and that sometimes it asks him not to test for certain viruses. I couldn't imagine why the CFIA would not want to test for certain viruses when one of its jobs is to protect Canadians from pathogens in food. The answer to this question came out during the evidentiary hearings about infectious salmon anemia (ISA), at the end of the Inquiry. The CFIA is obliged to report internationally what are known as reportable diseases to, for example, the World Health Organization (WHO). But if you don't find a reportable disease, you obviously don't have anything to report—and the legal requirements for testing are difficult to enforce. So the CFIA operates essentially as a trade organization. In order for Canadian products to be shipped and sold internationally, they must be free of positive tests involving reportable pathogens. And the best way to ensure that a product doesn't test positive for a pathogen is to not test the product for that pathogen in the first place. Or to only test at the minimally required level. Canada is not alone in its approach to testing. Many countries either don't test or do only minimal testing for reportable diseases. That way, their products can continue to be sold internationally. If Canada tested and reported all positive tests, it would put itself at a competitive disadvantage. So I get what this game is all about, but surely there has to be a way to balance everyone's and everything's needs and protect the fish at the same time?

The problem with the current approach is, of course, that many products aren't pathogen-free. And maybe the thinking is that nothing is truly pathogen-free. The rise in pathogens spreading around the world is testament to the role that global trade has played in the spreading of diseases. Certainly, governments everywhere are loath to require more testing for fear it would bring the global economy crashing down. However, passing virulent pathogens around the planet is clearly having a huge impact on wild fish populations. So how do we get past this? Truth be told, I'm not sure we can get past it. It's an international problem that is as old as time and will persist as long as there are jurisdictions that prioritize revenue over health and the reportable diseases rules remain effectively discretionary. With that in mind, it might be prudent for Canada to move as quickly as possible to closed systems for rearing farmed fish so that farmers can try to create products that are truly free of reportable diseases. I also hold out hope that the simple act of separating wild and farmed fish will be enough to allow nature to cleanse itself and our wild fish populations, including sockeye salmon, to return to their historical levels of abundance.

Does that make me an optimist? I suppose it does, but there are plenty of examples of nature rebounding when humans get out of the way. I believe the sockeye salmon can be one of them.

Cast of Characters

HONORABLE BRUCE COHEN:
Commissioner, The Uncertain Future of Fraser River Sockeye

LEONARD GILES:
Commission Registrar

TIMOTHY LEADEM:
Counsel, Conservation Coalition

LAURA RICHARDS:
Regional Director, Department of Fisheries and Oceans

WENDY BAKER:
Associate Commission Counsel

JIM WOODEY:
Former Pacific Salmon Commission Chief Biologist

BRIAN RIDDELL:
Former DFO Manager, Salmon Assessment

CARL WALTERS:
Professor Emeritus, the University of British Columbia

KEN WILSON:
Independent Fisheries Biologist

DAVID WELCH:
CEO, Kintama Research Services Limited

STEWART MCKINNELL:
Deputy Executive Secretary, North Pacific Marine Science

RICHARD BEAMISH:
Retired Marine Biologist, Department of Fisheries and Oceans

ALAN BLAIR:
Counsel, BC Salmon Farmers Association

GREGORY MCDADE, QC*:
Counsel, Aquaculture Coalition/Alexandra Morton

RANDALL PETERMAN:
Professor Emeritus, Canada Fisheries Risk Chair, Simon Fraser University

PATRICK MCGOWAN:
Associate Commission Counsel

SCOTT HINCH:
Professor, Pacific Salmon Ecology, University of British Columbia

KYLE GARVER:
Research Scientist, Department of Fisheries and Oceans

KRISTI MILLER:
Molecular Genetics, Department of Fisheries and Oceans

MITCHELL TAYLOR:
Senior Counsel, Government of Canada

KATHY L. GRANT:
Junior Commission Counsel

SONJA SAKSIDA:
Executive Director, Centre for Aquatic Health Sciences

SIMON JONES:
Research Scientist, Department of Fisheries and Oceans

CRAIG ORR:
Former Executive Director, Watershed Watch

BROCK MARTLAND:
Commission Staff Council, The Uncertain Future of Fraser River Sockeye

MICHAEL KENT:
Professor, Oregon State University

CRAIG STEPHEN:
Professor, Faculty of Veterinary Medicine, University of Calgary

CHRISTINE MACWILLIAMS:
Fish Health Veterinarian, Department of Fisheries and Oceans

Trial of an Iconic Species
—an illustrated screenplay

By
Scott Renyard

The Green Channel logo . . .

                              FADE TO:

Juggernaut Pictures logo . . .

                              FADE TO:

Title: Juggernaut Pictures presents . . .

                              FADE TO:

Title: A Scott Renyard documentary . . .

                              FADE TO:

INT. NEWS CONFERENCE — DAY

Commissioner Cohen is standing at the podium.

                COMMISSIONER COHEN
          As you all know, in November
          of 2009 I was appointed as
          commissioner to inquire into
          the decline of Fraser River
          sockeye salmon. The 2009
          sockeye return was the worst
          since the 1940s. And it was
          the third consecutive year
          without a commercial fishery.
          The sockeye fishery had been in
          a steady and profound decline
          since the early 1990s. What had
          caused this historic decline?
                    (MORE)

> COMMISSIONER COHEN (CONT'D)
> And what does the future hold
> for this iconic species? These
> were but a few of the many
> issues and questions that were
> sought answers.

FADE TO:

Title: Trial of an Iconic Species . . .

OVER BLACK — A heartbeat.

EXT. FRASER RIVER (1982) — DAY

A school of silver, slightly coloured sockeye
struggle against the rapids of the Fraser
Canyon near Boston Bar.

A flat, red line appears on the left side of
the frame and moves across to the right. It
overlays the struggle of the sockeye against
the current. Then it pulses, matching the
beating heart.

FADE TO:

GRAPH — of recruits per spawner and the year.
The heartbeat was mimicking the declining
return of sockeye spawners from the early
1990s. And the line goes flat.

FADE OUT:

CHAPTER 1

The Question of Science.

Or,

"Is the Fisheries department

focused on spin?"

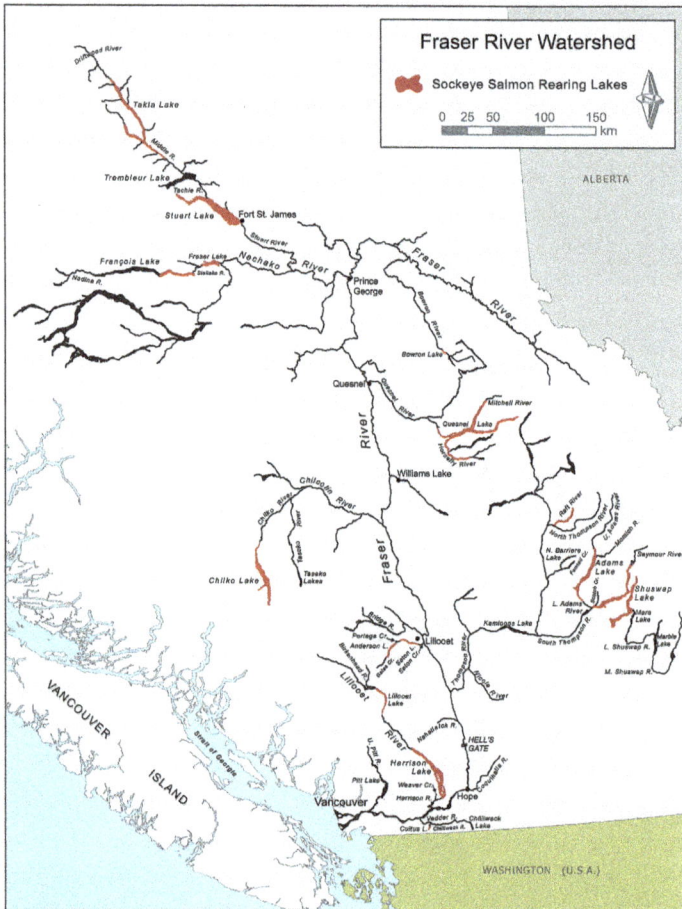

Fraser River Watershed

Sockeye Salmon Rearing Lakes

0 25 50 100 150 km

ALBERTA

Driftwood River
Takla Lake
Middle R.
Trembleur Lake
Tachie R.
Stuart Lake   Fort St. James
Stuart River
François Lake   Fraser Lake   Nechako   River
Nadina R.   Stellako R.
Prince George
Bowron River
Bowron Lake
Quesnel
Mitchell River
Quesnel   Lake
Horsefly River
Williams Lake
Chilcotin   River
Chilko River
Taseko River
Chilko Lake   Taseko Lakes
Bell River
North Thompson River
Upper Adams River
N. Barriere Lake
Seymour River
Adams Lake
Shuswap Lake
Mara Lake
Fraser
Seton R.
Portage Cr.
Anderson L.
Lillooet
L. Adams River
Kamloops Lake
L. Shuswap R.   Little Shuswap Lake
South Thompson R.
M. Shuswap R.
Lillooet   River
Lillooet Lake
Nahatlatch R.
HELL'S GATE
Coquihalla R.
VANCOUVER ISLAND
Strait of Georgia
Pitt Lake   Weaver Cr.
Harrison Lake
Harrison R.
Vancouver   Hope
Vedder R.   Chilliwack Lake
Cultus L.   Chehalis R.

WASHINGTON (U.S.A.)

Covering approximately 240,000 square kilometres, the Fraser River Watershed is British Columbia's largest watershed. The main stem of the Fraser River is 1,375 kilometres long, and numerous rivers and streams drain into it, including the Stuart, Nechako, Quesnel, Chilcotin, Thompson and Harrison Rivers. The Fraser River was once home to 32 distinct sockeye populations, but eight populations were either extinct or near extinct by 2010. The surviving 24 populations were the subject of the Cohen Inquiry. Most sockeye rear in freshwater lakes like the Shuswap, Quesnel, Stuart and Chilko. The notable exceptions are sockeye in the Harrison Lake system, which migrate quickly into the Salish Sea as fry.
(Map: Commission of Inquiry into the Decline of Sockeye Salmon in the Fraser River (Canada), The Uncertain Future of Fraser River Sockeye, Bruce I. Cohen, Commissioner.)

INT. COURTROOM — DAY

The Commissioner enters and takes a seat.

LOWER THIRD: Honorable Bruce Cohen, Commissioner.

> REGISTRAR GILES (O.S.)
> We are now resumed.

INT. COURTROOM — MOMENTS LATER

Everyone except Tim Leadem, counsel for the Conservation Coalition, takes their seats.

> MR. LEADEM
> For the record, Leadem, initial
> T., appearing as counsel for
> the Conservation Coalition.

LOWER THIRD: Tim Leadem, Senior Counsel, Conservation Coalition.

> MR. LEADEM
> I've spoken with you before,
> Dr. Richards.

INT. COURTROOM — MOMENTS LATER

Dr. Laura Richards, head of Science for DFO, glances down at her notes.

ON MR. LEADEM

                    MR. LEADEM
          You're the regional director
          and head of Science for the
          Pacific region as I understand
          it.

ON DR. RICHARDS

                    DR. RICHARDS
          Yes.

                    MR. LEADEM (O.S.)
          This is an email from Allison
          Webb to yourself.

ON DR. RICHARDS — looking down at her notes.

                    MR. LEADEM (O.S.)
          And Ms. Webb says, "This
          speech may be a little short
          now, but after speaking with
          Laura Richards, we had some
          strong concerns over some of
          the information contained which
          was not factually accurate so
          needed to delete it.

ON THE EMAIL — It slides into frame. Then
on the words "some information on Alexandra
Morton's website."

                    MR. LEADEM (O.S.)
          We have redrafted. Further
          Laura suggested that you may be
          able to find some information
          on Alexandra Morton's
          website . . .

BACK TO MR. LEADEM

                    MR. LEADEM
          . . . which points to some
          successes with respect to sea
          lice and pink salmon survival."
          So this is the speech that
          essentially deals with sea
          lice, is it not?

PAN TO DR. RICHARDS

                    DR. RICHARDS
          Ah, it could be, apparently,
          yeah.

**LOWER THIRD: Dr. Laura Richards, Regional
Director, Science, Department of Fisheries
and Oceans.**

                    MR. LEADEM (O.S.)
          If we could pull that, Mr.
          Lunn, 625, the speech that's
          attached,

Exhibit 625 — On the words "Speaking notes for a member of Parliament."

> MR. LEADEM (O.S.)
> . . . "Speaking notes for A
> Member of Parliament,"

ON DR. RICHARDS — looking at her monitor.

> MR. LEADEM (O.S.)
> . . . it talks about the
> aquaculture industry, and I see
> this paragraph:

ON THE DOCUMENT — finding the words "That is why the Government of Canada has invested $23.5 million over five years." Highlight "$23.5 million."

> MR. LEADEM (O.S.)
> . . . "That is why the
> Government of Canada has
> invested $23.5 million over
> five years as part of the . . .

PANNING OVER — to highlight "Aquaculture Innovation and Market Access Program."

> MR. LEADEM (O.S.)
> . . . Aquaculture Innovation
> and Market Access Program . . .

BACK TO MR. LEADEM

                    MR. LEADEM
          . . . to help establish a
          vibrant and sustainable
          Canadian aquaculture industry."
          Are those the kind of facts
          that you're double-checking?

PAN TO DR. RICHARDS

                    DR. RICHARDS
          No, they would not be.

                    MR. LEADEM (O.S.)
          That's because you're a doctor
          of science,

BACK TO MR. LEADEM

                    MR. LEADEM
          . . . not a "doctor of spin,"
          isn't that right?

                    DR. RICHARDS (O.S.)
          Well, that particular
          program . . .

ON DR. RICHARDS

                    DR. RICHARDS
          . . . was not a program that
          was dealt with through the
          Science program.

BACK TO MR. LEADEM

> MR. LEADEM
>
> So you're simply looking at these speeches from an aspect of, of commenting on the science.

> DR. RICHARDS (O.S.)
>
> That's correct.

> MR. LEADEM
>
> Is that fair to say?

> DR. RICHARDS (O.S.)
>
> Yes.

FADE TO:

> MR. LEADEM (O.S.)
>
> Do you recall this email?

ON DR. RICHARDS

> DR. RICHARDS
>
> Ah, I pay, I must say I do pay more attention to email which I'm directly sent than those that I'm just copied on. But I do recall that there was some question about what we could get in terms of various evidence from, from our own work on this particular topic.

THE COMMISSIONER — glances up at Richards.

                    MR. LEADEM (O.S.)
          All right.

PAN TO DR. RICHARDS

                    MR. LEADEM (O.S.)
          I'm curious about the first
          sentence: "Have we done any
          sampling that would . . .

ON MR. LEADEM

                    MR. LEADEM
          . . . counteract the findings
          of Alexandra's sockeye research
          near the Discovery Islands?"
          That would be Dr. Alexandra
          Morton's work, is that right?

                    DR. RICHARDS (O.S.)
          Ah, yes.

                    MR. LEADEM
          Is DFO in the business of
          trying to counteract the work
          of other scientists?

WHIP PAN TO DR. RICHARDS

                    DR. RICHARDS
          Ah, DFO is in the business of
          trying objectively to get to
          the truth.

ON MR. LEADEM

                    MR. LEADEM
          Can you explain why so much
          attention was being focused
          upon sea lice back then?

PAN TO DR. RICHARDS

                    DR. RICHARDS
          I think the attention that
          was being focused on sea
          lice was in response, ah, to
          other work that was being
          done, in response to things,
          other things that were being
          published, ah, in the media.
          And, ah, you know, that's why
          that was becoming apparent.

BACK TO MR. LEADEM

                    MR. LEADEM
          You, as a scientist, are then
          in the business of essentially
          responding to . . .

ON DR. RICHARDS

                    MR. LEADEM (O.S.)
          . . . media, letters, letters
          to the editor, letters that
          appear in the media.

BACK TO MR. LEADEM

                    MR. LEADEM
          And it troubles me, because
          when you attended earlier, some
          of the questions I put to you
          and to Dr. Mithani . . .

ON DR. RICHARDS

                    MR. LEADEM (O.S.)
          . . . specifically were that
          scientists should be careful
          to insulate themselves from
          becoming too political, from
          becoming protagonist too much.

ON MR. LEADEM

                    MR. LEADEM
          For either what their
          department was advocating
          or what their minister was
          advocating. Do you recall that
          general tenor of the discussion
          that I had with you and Dr.
          Mithani on that occasion?

                    DR. RICHARDS (O.S.)
          Ah, yes.

PAN TO DR. RICHARDS

                    DR. RICHARDS
          Yes, and I think I also need
          to correct exactly what you've
          said. It's not that I am
          writing, I am not writing, ah,
          letters to the media.

ON MR. LEADEM

                    DR. RICHARDS (O.S.)
          I am providing factual
          information that is going into
          those letters.

ON DR. RICHARDS

                    MR. LEADEM (O.S.)
          Do I think it's ok you agree
          with me that it's a very fine
          line between . . .

BACK TO MR. LEADEM

                    MR. LEADEM
          . . . vetting an email or
          vetting a potential media
          release from Department of
          Fisheries and Oceans for
          accuracy and actually spinning
          that in terms of how that is
          going to be portrayed in the
          media. There's a very fine line
          there, is it not?

                    DR. RICHARDS (O.S.)
     I don't dispute that there are,
     there, there are lines all over
     this work, yes . . .

PAN BACK TO DR. RICHARDS

                    DR. RICHARDS
     . . . but as I do want to
     emphasize is that it's
     important that for our
     perspective that science is
     seen to be objective, and
     we try to maintain that
     objectivity.

PANNING ACROSS — but not getting there.

                    MR. LEADEM (O.S.)
     Thank you. Those are my
     questions.

                              FADE OUT:

CHAPTER 2

The Question of Over Escapement.

Or,

"Is it bad to underfish?"

Sockeye, like all Pacific salmon species, spawn only once and then die. Female spawners produce between 3,000 and 4,000 eggs. If a female lays 3,000 eggs, only 14%—or 420—survive to the fry stage. Of those 420, only 29%—or 122—survive the fry stage to become smolts; and of those 122, 91%—or 110—will die before reaching adulthood. In other words, out of 3,000 eggs, only 12 will become reproductive adults that could return to spawn. The average harvest rate is about 70%, which leaves just four fish to spawn and reproduce the next cohort of sockeye. (Source: Commission of Inquiry into the Decline of Sockeye Salmon in the Fraser River (Canada), The Uncertain Future of Fraser River Sockeye, Bruce I. Cohen, Commissioner. Photo: Anissa Reed.)

INT. COURTROOM — MORNING

SUPERSCRIPT: February 9, 2011 . . .

The Commissioner bows and takes his seat.

> REGISTRAR GILES (O.S.)
> The hearing is now resumed.

ON COMMISSIONER COHEN — He smiles in the direction of Wendy Baker, indicating she can proceed.

ON A MONITOR — A smiling Dr. Carl Walters appears via Skype.

> MS. BAKER (O.S.)
> Thank you. Good morning, Mr. Commissioner. Today we have Dr. Carl Walters in Florida on the screen, looming over us here larger than life.

CLOSE ON MS. BAKER

LOWER THIRD: Ms. Wendy Baker, Associate Commission Counsel.

> MS. BAKER
> And, ah, we have Mr. Ken Wilson, who you met the last two days,

PANNING — across the panellists.

                    MS. BAKER (O.S.)
        . . . Dr. Jim Woodey, in the
        centre on the panel, and Dr.
        Brian Riddell.

ON COMMISSIONER COHEN — making notes.

                    MS. BAKER (O.S.)
        Now, Dr. Woodey, just to lead
        off on this notion of cyclic
        dominance, can you just give us
        some help on that?

ON DR. WOODEY

**LOWER THIRD: Dr. Jim Woodey, Former Pacific
Salmon Commission Chief Biologist.**

                    DR. WOODEY
        Cyclic dominance involves one
        large return year,

ON COMMISSIONER COHEN — listening to the
testimony.

**POP-UP — Cyclic Dominance = An apparent
influence of a large run of sockeye on a
four-year cycle**

                    DR. WOODEY (O.S.)
        . . . generally a subdominant
        line year, and then two years
        where . . .

BACK TO DR. WOODEY

                    DR. WOODEY
          . . . the abundance is somewhat
          lower . . .

ON MR. WILSON

                    DR. WOODEY (O.S.)
          . . . to a . . .

BACK TO DR. WOODEY

                    DR. WOODEY
          . . . few percent of the
          dominant year abundance.

ON MR. RIDDELL

**LOWER THIRD: Dr. Brian Riddell, Former DFO
manager, Salmon Assessment.**

                    DR. RIDDELL
          We're still trying to
          understand what it is and that.
          But I think that the notion
          that it's maintained by fishing
          is not accepted.

ON MS. BAKER

                    DR. RIDDELL (O.S.)
          We do know that it's not as
          simple as just food production
          between years, right, because
          we do have data showing . . .

BACK TO MR. RIDDELL

              DR. RIDDELL
. . . that the recovery of the
lake is certainly sufficient to
produce food far in excess of
what would be required by the
small number of fish . . .

ON DR. WALTERS

              DR. RIDDELL (O.S.)
. . . in the subdominant cycles.

              MS. BAKER (O.S.)
Ah, Dr. Walters?

**LOWER THIRD: Dr. Carl Walters, Professor
Emeritus, the University of British Columbia.**

              DR. WALTERS
I think we can say pretty
definitely that it was . . .

BACK TO COMMISSIONER COHEN

              DR. WALTERS (O.S.)
. . . thought the stocks were
exhibiting violent cycles . . .

BACK TO DR. WALTERS

> DR. WALTERS
> . . . before the fisheries
> became large enough to cause
> those cycles.

PULLING BACK — the Commissioner comes into frame.

> DR. WALTERS
> That's one of the really
> important findings from the
> Gilhousen work, that there was
> a, a cyclic pattern established
> by the early 1890s.

ON MR. WILSON

LOWER THIRD: Ken Wilson, Independent Fisheries Biologist.

> MR. WILSON
> I would like to make a point.
> You know, we go back to pre-
> contact and have a discussion
> about what salmon populations
> might have been like. Ah, I
> think we can,

EXT. ADAM'S RIVER — DAY

Silver sockeye school in the back eddy.

>   MR. WILSON (O.S.)
>   . . . most of us agree, that
>   populations on average were
>   larger and, ah, escapements
>   at some times were very
>   substantial.

INT. COURTROOM — DAY

Mr. Wilson is back on screen.

>   MR. WILSON
>   Um. We have, you know, 40
>   million fish perhaps, but
>   I've got a quote here from Dr.
>   Ricker . . .

DOCUMENT — Scroll down and move tight to find
the number 160 million.

>   MR. WILSON (O.S.)
>   . . . that peak abundance in
>   Fraser sockeye might be as high
>   as 160 million. That was quoted
>   by Northcote and Atagi. Over
>   escapement really can only be
>   understood if we call it by
>   its . . .

ON DR. WOODEY

>   MR. WILSON (O.S.)
>   . . . proper name, and . . .

BACK TO MR. WILSON

>MR. WILSON
>. . . I think in this case,
>it's underfishing. We're not
>harvesting all the fish that
>have been identified as surplus
>to the escapement goal using
>the kinds of management models
>and processes we currently use.

DR. RIDDELL — listening.

>MR. WILSON (O.S.)
>I suspect we saw a different
>distribution of spawners,

BACK TO MR. WILSON

>MR. WILSON
>. . . ah, very large
>escapements from freshwater
>areas,

INT. SHUSWAP LAKE — AFTERNOON

A school of sockeye fry swim past a carcass.

>MR. WILSON (O.S.)
>. . . lots of carcasses and
>nutrients that benefited, in
>all likelihood, large sections
>of the Fraser watershed.

ON MS. BAKER

>   MR. WILSON (O.S.)
> So I guess, in sum, I'm simply
> suggesting that, ah . . .

BACK TO MR. WILSON

>   MR. WILSON
> . . . underfishing — I know
> when I started with the
> department as a biologist, it
> was really the only thing a
> management biologist could
> do to get himself in serious
> trouble was to underfish.

FADE TO:

INT. COURTROOM — DAY

Mr. Leadem is at the microphone.

>   MR. LEADEM
> Could or should fisheries
> managers try to iron out the
> peaks and valleys so that we
> can arrive at more of a level
> aspect of, of return year after
> year? And maybe I will start
> with you, Dr. Riddell?

ON MR. RIDDELL

                    DR. RIDDELL
          You can probably manage to
          maximize production, but you
          should let the fish choose
          how they're actually going to
          restore production.

ON THE MONITOR — Dr. Walters is on the
screen.

                    MR. LEADEM (O.S.)
          Dr. Walters?

                    DR. WALTERS
          Around the Pacific Rim there
          have been four major cases
          where cyclic-dominant stocks
          have, either deliberately or
          inadvertently, had their cyclic
          dominance break down. The first
          of those was the Kvichak stock
          in . . .

MAP — of the west coast. Highlight Kvichak
River location.

                    DR. WALTERS (O.S.)
          . . . Alaska, in Bristol Bay.
          It was the world's largest
          salmon stock.

ON DR. WALTERS

                    DR. WALTERS
          And as a more or less
          deliberate experiment, they
          broke up its very violent
          cycle. It is now listed as a
          stock of concern in Alaska. The
          second case was Rivers Inlet,
          here in BC,

BACK TO MAP — Rivers Inlet.

                    DR. WALTERS (O.S.)
          . . . where we deliberately
          shut down the fishery and tried
          to rebuild the stock to test
          for having the, ah, having the
          benefits of higher spawning
          runs.

BACK TO THE MONITOR — Dr. Walters is checking
his notes.

                    DR. WALTERS
          Shortly after that experiment
          started, Rivers collapsed and
          it still hasn't recovered.

BACK TO MAP — Stuart and Quesnel Lakes are
highlighted in red.

                    DR. WALTERS (O.S.)
          Then we have the Late Stuart
          and Quesnel stocks . . .

ON MR. LEADEM — listening.

                    DR. WALTERS
          . . . on the Fraser, strongly
          cyclic-dominant, but the cycle
          has broken down over the last
          15 years,

BACK TO MR. WALTERS

                    DR. WALTERS
          . . . and in both cases
          the mean productivities of
          those stocks have dropped
          dramatically. So I think
          the evidence is pretty clear
          that it isn't a good idea,
          that there's something in
          the biology of these cyclic
          stocks that makes them very
          unproductive when they're, ah,
          when they're not in a cyclic
          mode.

                              FADE OUT:

CHAPTER 3

The Question of the
Marine Environment.
Or,
"Why are young sockeye
disappearing in the ocean?"

Evidence presented at the Cohen Inquiry suggested that early marine survival of sockeye smolts was very poor and that most of the recent unusual mortality occurred north of the Salish Sea. The narrow channels north of the Salish Sea are where most of BC's open net pen fish farms are located. (Photo: Scott Renyard)

OVER BLACK

          REGISTRAR GILES (O.S.)
    Order!

INT. COURTROOM — DAY

SUPERSCRIPT: July 6, 2011 . . .

The Commissioner enters and bows. He exchanges a few words with John Lund, an Inquiry staff member.

PANNING ACROSS — from Giles to Baker.

          REGISTRAR GILES
    The hearing is now resumed.

        COMMISSIONER COHEN (O.S.)
    Good morning, Ms. Baker.

           MS. BAKER
    Good morning.

WIDE — The Commissioner looks at the three panellists.

         MS. BAKER (O.S.)
    We have with us today three
    doctors. We have Dr. Beamish,
    Dr. Welch and Dr. McKinnell.

PAN TO DR. WELCH — He looks at his monitor.

                    MS. BAKER (O.S.)
          I'd like to move, now, to Dr.
          Welch. All right,

BACK TO MS. BAKER

                    MS. BAKER
          . . . so this work that you've
          just described showed the
          Lower Fraser, Mission to the
          mouth of the Fraser, as having
          reasonably good survival. What
          do you know about survival from
          the lake to Mission?

INT. CULTUS LAKE — DAY

A school of sockeye fry swim by.

                    DR. WELCH (O.S.)
          In Cultus Lake, most of the
          mortality in the Fraser
          River . . .

EXT. COUNTING FENCE — DAY

A number of sockeye smolts are in the trap.

                    DR. WELCH (O.S.)
          . . . that we measured occurred
          between release and . . .

INT. CREEK — DAY

Sockeye smolts swim in a school downstream.

LOWER THIRD: Dr. David Welch, CEO, Kintama Research Services Limited.

                    DR. WELCH
          . . . essentially the exit from
          Sweltzer Creek into the main
          stem of the Fraser. It's a
          small, trans — clear . . .

EXT. SWELTZER CREEK — DAY

Establish.

ON A SIGN — "Cultus Lake Sockeye at Risk"

                    DR. WELCH (O.S.)
          . . . river and it looks like
          a lot of things do eat sockeye
          within that, so there's,
          there's predators there.

BACK TO. DR. WELCH

                    DR. WELCH
          In 2010, we observed something
          very similar up in Chilko Lake.

MAP — Highlight Chilko Lake and its tributary
to the Fraser River.

                    DR. WELCH (O.S.)
          The first section of the
          migration, a clear water river
          running from Chilko Lake, we
          had much higher mortality
          there.

The safe zone of the Fraser River is lit up
in green.

                    DR. WELCH (O.S.)
          And then once they reached the
          main Fraser River, mortality
          rates dropped.

ON MS. BAKER

                    MS. BAKER
          Do you have any views on the
          importance of the Strait of
          Georgia . . .

SOUTH COAST MAP — The Strait of Georgia
lights up in bright blue.

                    MS. BAKER (O.S.)
          . . . in relation to returns
          of adult sockeye? Fraser River
          sockeye?

EXT. FRASER RIVER — DAY

The clear Vedder River water is mixing with
the muddy Fraser River water.

                    MS. BAKER (O.S.)
          And can you just describe what
          the results of that work have
          shown?

                    DR. WELCH (O.S.)
          Overall, the issue is that in
          the Lower Fraser River . . .

ON DR. WELCH

                    DR. WELCH
          . . . the survival is high for
          all of the stocks . . .

MAP — tighter on the Strait of Georgia.

                    DR. WELCH (O.S.)
          . . . or all of the years, and
          then in the northern Strait of
          Georgia,

Strait of Georgia lights up in green.

                    DR. WELCH (O.S.)
          . . . which is from the Fraser
          River mouth to the north end of
          Texada Island, Comox to Powell
          River, survival is high and,
          um, fairly stable.

BACK TO DR. WELCH

> DR. WELCH
> Ah, and then the surprise for
> us in 2010 . . .

PANNING MAP — to include the area north of
Vancouver Island and Queen Charlotte Sound.

> DR. WELCH (O.S.)
> . . . is that the survival from
> the north end of Texada Island
> to the, near the exit of Queen
> Charlotte Strait,

Johnstone Strait and Queen Charlotte Strait
go red.

> DR. WELCH (O.S.)
> . . . survival was only about a
> third to a quarter in 2010 for
> these smolts.

> FADE TO:

ON DR. MCKINNELL — looking at Ms. Baker.

> MS. BAKER (O.S.)
> Dr. McKinnell,

DOCUMENT — Exhibit 1291, Cohen Commission
Technical Report 4 — Marine Ecology — Feb
2011.

MS. BAKER (O.S.)
. . . you have prepared a
report for the Commission of
Inquiry which is described
as . . .

ON MS. BAKER

MS. BAKER
. . . "Technical Report 4, The
Decline of Fraser River Sockeye
Salmon . . .

ON A PAGE — The executive summary. It shows
two main questions.

MS. BAKER (O.S.)
. . . in Relation to Marine
Ecology."

DR. MCKINNELL (O.S.)
Yes.

ON DR. MCKINNELL

MS. BAKER (O.S.)
In this report, you were asked
to . . .

BACK TO — Exhibit 1291, Executive Summary.
The two questions are highlighted.

MS. BAKER (O.S.)
. . . answer two questions, the
first being:

**PLATE — Typing the words "Do marine
conditions explain the 2009 decline?" and
"Is there evidence of declines in marine
productivity?"**

MS. BAKER (O.S.)
. . . Do marine conditions
explain the 2009 decline? And
secondly, is there evidence of
declines in marine productivity
or changes in Fraser River
sockeye distribution? Those
were the two big questions you
were asked.

BACK TO DR. MCKINNELL

DR. MCKINNELL
Yes.

PANNING OVER TO MS. BAKER

MS. BAKER
You've talked already about the
number of extremes that you saw
in various areas in 2007. Are
those extremes relevant to the
analysis of the cause of low
returns in 2009?

LOWER THIRD: Dr. Stewart McKinnell, Deputy Executive Secretary, North Pacific Marine Science.

> DR. MCKINNELL
> I think one of the things that
> I have to do at this point is,
> is point out that it's not just
> the low returns in 2009 that
> have to be satisfied. There are
> a collection of observations
> that go along with this that
> also have to be explained.

MAP — The Columbia River lights up in green.

> DR. MCKINNELL (O.S.)
> In that same year we had double
> the average returns of sockeye
> to the Columbia River.

PANNING MAP — to the Barkley Sound area. It lights up in green.

> DR. MCKINNELL (O.S.)
> We had better than expected
> returns of sockeye to Barkley
> Sound. We had very low returns
> of the one-year-old smolts from
> most populations that entered,
> the, ah, the Strait of Georgia,
> and that's the point you've
> raised about the 2009.

BACK TO MAP — The Strait of Georgia is green. Then, the Harrison River watershed.

> DR. MCKINNELL (O.S.)
> But we had record high returns from sockeye that were in the Strait of Georgia in that same summer to Harrison River.

BACK TO DR. MCKINNELL

> DR. MCKINNELL
> And in fact, I think the two-year-old smolts from Chilko Lake survivals were, ah, not affected — measured at Chilko Lake were not affected as the one-year-old smolts. And we have Dr. Welch's observation of typically, typical survival of his acoustically tagged Cultus Lake sockeye through, through the Georgia Strait in 2007. So what one needs to do is develop a model that, that somehow satisfies all of these concurrent observations.

ON MS. BAKER — for a moment.

                    DR. MCKINNELL
          Certainly placing the mortality
          of this brood year, of the
          2007 age-one smolts in Queen
          Charlotte Strait/Queen
          Charlotte Sound, has the
          possibility to satisfy all of
          these observations . . .

INT. OCEAN — DAY

A school of young sockeye salmon in a net.

                    DR. MCKINNELL (O.S.)
          . . . because the age-one
          smolts from — and two-year-old
          smolts from the Fraser River
          must,

ON DR. MCKINNELL

                    DR. MCKINNELL
          . . . that's just their nature,
          is they generally swim through
          Johnstone Strait and Queen
          Charlotte Strait and Queen
          Charlotte Sound.

                    MS. BAKER (O.S.)
          But they would also go through
          the Strait of Georgia, so why
          would you eliminate that?

                    DR. MCKINNELL
          Ah, the only reason I would
          — I mean, I'm not eliminating
          it. Ah, I just would not put
          so much emphasis on it, as the
          site,

PANNING TO DR. BEAMISH — He nods in response
to Dr. McKinnell's assertion and makes a
note.

                    DR. MCKINNELL (O.S.)
          . . . simply because of what
          I've explained before, that I
          was looking for extremes in
          physics and in, potentially
          in chemistry that would, that
          would — and where those
          occurred.

TO DR. WELCH

                    MS. BAKER (O.S.)
          Do any of the conditions
          that you've reviewed with us
          today . . .

BACK TO MS. BAKER

                    MS. BAKER
          . . . indicate a causation . . .

ON THE REPORT — panning the extensive list of
topics reviewed.

> MS. BAKER (O.S.)
> . . . between those conditions
> and Fraser River sockeye
> survival?

BACK TO DR. MCKINNELL

> DR. MCKINNELL
> No.

FADE TO:

ON DR. BEAMISH

> MS. BAKER (O.S.)
> I'd like to move to Dr. Beamish
> now.

ON THE COMMISSIONER'S MONITOR — Exhibit 1309
is on the screen.

> MS. BAKER (O.S.)
> Can you just give us a very
> brief overview of this report?

ON DR. BEAMISH

**LOWER THIRD: Dr. Richard Beamish, Retired
Marine Biologist, The Department of Fisheries
and Oceans.**

> DR. BEAMISH
> We've done a number of surveys
> in the Strait of Georgia . . .

STRAIT OF GEORGIA MAP — shows the location of his trawl location sites.

> DR. BEAMISH (O.S.)
> . . . around 1,800 sets that we've made over that period of time.

INT. SALISH SEA — DAY

Juvenile salmon swim and feed near the surface.

> DR. BEAMISH (O.S.)
> Almost 98 percent of the fish that we catch in the surface 30 metres are juvenile herring or juvenile salmon.

ON MS. BAKER

> DR. BEAMISH (O.S.)
> Now, all of those fish in 2007 . . .

BACK TO DR. BEAMISH

> DR. BEAMISH
> . . . ended up having poor growth or . . .

INT. OCEAN — DAY

More juvenile herring swim along feeding.

                    DR. BEAMISH (O.S.)
          . . . poor survival or both.

INT. COURTROOM — MOMENTS LATER

Dr. Beamish speaks into the mic.

                    DR. BEAMISH
          And perhaps the most
          spectacular, um, um,
          observation was with juvenile
          herring.

Commissioner Cohen makes a note.

                    DR. BEAMISH (O.S.)
          And in 2007, their survey
          estimates . . .

BACK TO DR. BEAMISH

                    DR. BEAMISH
          . . . indicated that they
          had . . .

BACK TO THE HERRING — underwater.

                    DR. BEAMISH (O.S.)
          . . . the lowest abundance
          of juvenile herring in their
          history. In addition to that,
          then when the herring that were
          spawned in that year . . .

BACK TO DR. BEAMISH

                    DR. BEAMISH
          . . . were recruited into the
          fishery in 2010, the commercial
          fishery, those recruits usually
          represent about 60 percent
          of the um, of the commercial
          catch.

THE HERRING — are spooked.

                    DR. BEAMISH (O.S.)
          And in 2010,

ON DR. BEAMISH

                    DR. BEAMISH
          . . . if I remember correctly,
          it was around six percent. It
          was the lowest recruitment ever
          recorded of a year class.

                                        FADE TO:

INT. COURTROOM — DAY

The Commissioner is at his desk.

SUPERSCRIPT: July 7, 2011 . . .

INT. COURTROOM — MOMENTS LATER

Mr. Blair is at a microphone.

                    MR. BLAIR
          Mr. Commissioner, for the
          record, Alan Blair, counsel
          for the BC Salmon Farmers
          Association.

ON DR. BEAMISH

                    MR. BLAIR (O.S.)
          Yesterday I heard Dr. Beamish
          indicate that . . .

BACK TO MR. BLAIR

**LOWER THIRD: Alan Blair, Senior Counsel, BC
Salmon Farmers Association.**

                    MR. BLAIR
          . . . in his work, in
          particular I'm referring to the
          summer of 2007.

BACK TO DR. BEAMISH

                    MR. BLAIR (O.S.)
          He sampled fish and across a
          wide spectrum of species,

ON MR. BLAIR

                    MR. BLAIR
          . . . herring as well as a
          number of different species of
          salmon. He found fish with
                    (MORE)

                    MR. BLAIR (CONT'D)
          empty bellies. He found fish
          that were stressed from lack of
          food.

TO DR. WELCH

                    MR. BLAIR (O.S.)
          My premise to the three of you
          is that I don't think anything
          that I heard . . .

ON MR. BLAIR

                    MR. BLAIR
          . . . Dr. McKinnell say
          yesterday was contradicted
          by Dr. Beamish, and I didn't
          think I heard anything that
          Dr. Beamish said yesterday
          with respect to his trawls
          in the Strait of Georgia was
          contradicted by Dr. McKinnell.

REVERSING TO DR. WELCH

                    DR. WELCH
          The broader issue that I, I
          take issue with . . .

DOCUMENT — displaying data collected as part
of the Beamish trawls.

                    DR. WELCH (O.S.)
          . . . is not Dr. Beamish's
          excellent data,

BACK TO DR. WELCH

                    DR. WELCH
          . . . but the inference that we
          know that the survival problem,
          with very high likelihood,
          happened in the Strait of
          Georgia. And that's not
          reasonable given the data.

ON COMMISSIONER COHEN

                    COMMISSIONER COHEN
          We'll take a break.

                                        FADE TO:

Mr. Gregory McDade, QC, is at the microphone
and picks up his glasses to read his notes.

                    MR. MCDADE
          My name is Gregory McDade, and
          I am counsel for Dr. Morton and
          the Aquaculture Coalition. I'll
          have a few questions for you.
          Let me start first with you,
          Dr. Beamish.

ON DR. BEAMISH — He appears uncomfortable.

MR. MCDADE (O.S.)
As I understood from your
evidence yesterday,

BACK ON MR. MCDADE

MR. MCDADE
. . . ah, you formed the
conclusion, ah, either before
2009 returns were in or very
shortly thereafter, that
there was a problem with prey
abundance in the Strait of
Georgia. And you then went and
recruited Dr. Thomson . . .

ON DR. BEAMISH

MR. MCDADE (O.S.)
. . . to try and answer the
question of why. Is that fair
enough?

DR. BEAMISH
Yes.

MR. MCDADE (O.S.)
Before getting any of this wind
and salinity and his . . .

ON DR. MCKINNELL

MR. MCDADE (O.S.)
. . . MLD modelling done,

POP-UP — MLD Modelling: Mixed Layer Depth Modelling

ON DR. BEAMISH

>   MR. MCDADE (O.S.)
>   . . . you'd already formed the
>   conclusion that, that prey
>   abundance was the issue.

ON MR. MCDADE — briefly.

>   DR. BEAMISH (O.S.)
>   It was . . .

Mr. McDade glances at Dr. Beamish.

BACK TO DR. BEAMISH

>   DR. BEAMISH
>   . . . it was a possibility. I
>   don't think I would say that I
>   had finalized the conclusion,
>   no.

ON COMMISSIONER COHEN — making notes.

>   MR. MCDADE (O.S.)
>   Well, all I'm trying to
>   establish is the process,

DR. BEAMISH — listening.

                    MR. MCDADE (O.S.)
          . . . the scientific process.
          What I think I understand here
          is that you had come up with
          the idea of poor prey abundance
          first, and then you were trying
          to establish a model that would
          be, ah, consistent with your
          data that would confirm that.

DR. BEAMISH — answers.

                    DR. BEAMISH
          I can, I can only keep
          answering the question the
          same way. I mean, this could
          have been a disease, right?
          This could have been some
          catastrophic event that was,
          was unprecedented. And I
          used to say to people, "You
          know, maybe it's aliens." Now,
          obviously, I don't believe it's
          aliens, but the point is that
          something very anomalous was
          happening and we were trying to
          understand what it was. Winds
          was one of the components that
          we were looking at, and, and
          Dr. Thomson was a colleague who
          had very good wind data.

                              FADE OUT:

CHAPTER 4

The Question of Missing

Fraser River Spawners.

Or,

"Is this a much bigger problem?"

The rotting carcass of an adult sockeye in Shuswap Lake serves its last purpose as a nutrient boost in the lake where the next generation of sockeye fry will grow and rear for a year. The Adams River stock has experienced dramatic declines in abundance, as have most Fraser River stocks since the early 1990s. The Cohen Inquiry investigated whether sockeye declines in the Fraser River were unique in the eastern Pacific. With just a few exceptions, sockeye populations have actually been declining in a similar pattern up and down North America's Pacific coast. (Photo: Scott Renyard)

INT. COURTROOM — MORNING

> REGISTRAR GILES
> Order!

Commissioner Cohen enters.

> REGISTRAR GILES
> We are now resumed.

SUPERSCRIPT: April 20, 2011 . . .

ON THE COMMISSION COUNSEL PODIUM — Ms. Baker is at her podium.

> MS. BAKER
> Good morning, Mr. Commissioner.
> Today we have the, ah, Project
> 10 being tendered . . .

ON DR. PETERMAN — Establish.

> MS. BAKER (O.S.)
> . . . and Dr. Randall Peterman
> and . . .

ON DR. DORNER — She's appearing via Skype.

> MS. BAKER (O.S.)
> . . . Dr. Brigitte Dorner are
> here to testify . . .

TIGHT ON DR. PETERMAN — looking at his monitor.

                    MS. BAKER (O.S.)
          . . . for this analysis,

DOCUMENT — Title page of Exhibit 417, "Does
Over-Escapement Cause Salmon Stock Collapse?"

                    MS. BAKER (O.S.)
          . . . a paper which was
          discussed by Drs. Walters and
          Riddell earlier this year. That
          paper is referenced in your
          first paragraph.

BACK TO DR. PETERMAN

                    MS. BAKER (O.S.)
          How did your work relate to the
          work that was done earlier in
          Exhibit 417?

WIDER

                    DR. PETERMAN
          We found that the adult
          recruits came back in fewer
          numbers than the number of
          spawners, that is, below
          replacement, in only seven
          percent of the years . . .

FRASER WATERSHED MAP — showing the 19 stocks
of sockeye.

                    DR. PETERMAN (O.S.)
          . . . across all the 19 Fraser
          River stocks, all across the
          50 years approximately of the
          data.

Dr. Peterman checks his notes.

DOCUMENT — Panning the list of Fraser River
populations.

                    MS. BAKER (O.S.)
          Out of all the stocks that you
          reviewed, did any individual
          stocks show a relationship
          between . . .

ON MS. BAKER

                    MS. BAKER
          . . . high spawner abundance
          and low recruits?

ON DR. PETERMAN

                    DR. PETERMAN
          What we found was that
          the . . .

**POP-UP — Larkin Hypothesis suggests spawning
density affects future returns**

                    DR. PETERMAN
          . . . Larkin delayed density-
          dependence hypothesis only
          appears relevant to the . . .

MAP — The Quesnel stock lights up.

                    DR. PETERMAN (O.S.)
          . . . Quesnel sockeye stock in
          the Fraser.

COMMISSIONER COHEN — making a note.

                    DR. PETERMAN (O.S.)
          So that means that there is
          some factor or factors . . .

BACK TO DR. PETERMAN

                    DR. PETERMAN
          . . . other than spawner
          abundance in any of the years
          that's causing the productivity
          to go down. Because the Larkin
          model takes into account the
          effect of spawner abundance
          across generations, but it
          still shows a decrease in
          productivity in the last decade
          or so. So that means something
          else is causing that decline.

EXHIBIT — On the cover, turn the page to
reveal . . .

NEW PAGE — Show next section: "Comparison
of productivity patterns across sockeye
populations."

                MS. BAKER (O.S.)
          Moving to this next section:
          "Comparison of productivity
          patterns across populations."

BACK TO MS. BAKER

                MS. BAKER
          And asking if you could just
          state what the purpose of this
          analysis was.

ON DR. PETERMAN

                DR. PETERMAN
          First of all, we wanted to
          accurately describe these
          trends in Fraser River sockeye
          productivity that was the
          genesis . . .

ON MS. BAKER — checking her notes.

                DR. PETERMAN (O.S.)
          . . . of this Commission, so
          that we get a clearer picture.

ON DR. PETERMAN — He is animated, gesturing
as he frames the overall scale of the
decline.

                    DR. PETERMAN
          Not just what's in the news,
          but just really describe what
          has happened to the Fraser
          sockeye. The second, we also
          wanted to determine whether
          the decreasing trend in the
          Fraser sockeye abundance and
          productivity was shared by
          other sockeye populations on
          the west coast. In essence, we
          wanted to ask, is this a unique
          phenomenon to the Fraser, or is
          it widespread, and if it's, if
          it's not unique, how widespread
          is it? And the third purpose
          was . . .

BACK TO MS. BAKER

                    DR. PETERMAN (O.S.)
          . . . to provide to other
          researchers that are working on
          the various hypotheses . . .

BACK TO DR. PETERMAN

                    DR. PETERMAN
          . . . for the Commission in the
          12 projects that are going on.

GRAPHS — of shared trends grids over three
time periods.

                    MS. BAKER (O.S.)
          All right. So what do the
          shared productivity patterns
          that we see in your time
          trends tell us about causal
          mechanisms? Anything?

Dr. Peterman looks toward the Commissioner.

                    DR. PETERMAN
          To us it seems like there's
          a much greater chance that
          there's some shared trend
          across these populations to
          varying extents . . .

WEST COAST MAP — The 64 sockeye populations
are represented by red dots that pop up along
the coast.

                    DR. PETERMAN (O.S.)
          . . . causing a downward trend
          in productivity of all these
          stocks simultaneously.

PULLING BACK — The red dots disappear and the
focus is now on the water which lights up to
reveal the area occupied by Pacific sockeye
populations with the label "Area Occupied by
Pacific Sockeye."

                    DR. PETERMAN (O.S.)
          It should be a phenomenon
          that's got a large spatial
          scale.

BACK TO DR. PETERMAN

                    DR. PETERMAN
          Okay. So, that's the first
          hypothesis I think we can, if
          not rule out, put at a very low
          probability, is the freshwater
          events. Second, we can probably
          rule out delayed density-
          dependence as the shared source
          of downward-driving trend in
          productivity. Like I said
          before, it's definitely true
          for the Quesnel stock that
          delayed density-dependence
          seems to have occurred, but it
          does not seem to have played
          an important role in any of the
          other stocks we've looked at,
          and we fit the Larkin model, by
          the way, to all 64 populations,
          not just the Fraser. So I think
          we can rule out the delayed
          density-dependence argument for
          explaining this shared time
          trend.

EXT. HARRISON RIVER — DAY

Dead fish float in the river near Harrison
Lake.

                    DR. PETERMAN (O.S.)
          The third hypothesis that we
          can rule out is this en route
          mortality. So, as we were
          discussing earlier, the en
          route mortality is what happens
          to the adults . . .

BACK TO DR. PETERMAN

                    DR. PETERMAN
          . . . as they enter the river
          system and migrate up towards
          their spawning grounds.

INT. HARRISON RIVER — DAY

A silver, dead sockeye with lesions floats
by.

                    DR. PETERMAN (O.S.)
          So I think those are the
          only . . .

BACK TO DR. PETERMAN

                    DR. PETERMAN
          . . . three hypotheses that we
          can really rule out.

                                    FADE OUT:

CHAPTER 5

The Question of

Climate Change.

Or,

"Is the Fraser River

water too hot?"

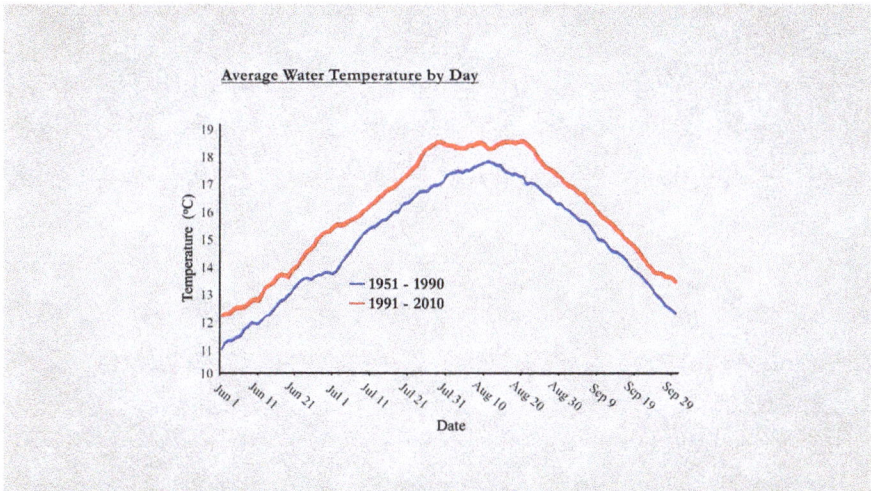

Average Water Temperature by Day

This graph from Exhibit 553 compares the temperature of Fraser River water from 1951–1990 to 1991–2010. The researchers discovered that Fraser River water experienced an increase in temperature of one degree from June to the end of September. The four main run timings of Fraser sockeye are displayed above the graph. One sockeye timing identified as the Lates since 1995 has been entering the river much earlier than usual. Higher water temperatures would presumably be more difficult for this group to adapt to since they traditionally enter the river in the fall, when water temperatures are cooler. Early entry by this group was likely because a large portion of these fish were sick and were trying to spawn before they died. (Source: Commission of Inquiry into the Decline of Sockeye Salmon in the Fraser River (Canada), The Uncertain Future of Fraser River Sockeye, Bruce I. Cohen, Commissioner.)

INT. COURTROOM — DAY

Establish. The Commissioner is already seated.

SUPERSCRIPT: March 8, 2011 . . .

> PATRICK MCGOWAN (O.S.)
> Good morning, Mr. Commissioner.

TIGHTER — Patrick McGowan is at the microphone.

> PATRICK MCGOWAN
> We have here today Dr. Scott Hinch and Eduardo Martins, the two authors . . .

DOCUMENT — Report 9, cover page. "A Review of Potential Change Effects on Survival of Fraser River Sockeye Salmon and an Analysis of Interannual Trends in En Route Loss and Pre-spawn Mortality."

> PATRICK MCGOWAN (O.S.)
> . . . of the report, who are here to give evidence to you on its contents. I'm wondering if you could briefly explain for the Commissioner,

BACK TO MR. MCGOWAN

LOWER THIRD: Patrick McGowan, Associate Commission Counsel.

                    PATRICK MCGOWAN
          . . . the phenomenon of climate
          change and how it is impacting
          on the Fraser River?

ON THE COMMISSIONER — He has the report title
on his monitor and looks at Dr. Hinch.

                    DR. HINCH (O.S.)
          Okay. Well, there's really
          three components to climate
          change.

EXT. MOUNTAIN — DAY

The trees are blanketed in haze.

                    DR. HINCH (O.S.)
          First, is a global issue
          dealing . . .

EXT. NORTH VANCOUVER SKY — DAY

A deep haze almost blocks the sun.

                    DR. HINCH (O.S.)
          . . . with greenhouse gas
          emissions and the increase
          we've seen in those in the last
          several decades.

EXT. BURRARD INLET — DAY

A tugboat crosses the inlet and the haze
blocks everything else from view.

                    DR. HINCH (O.S.)
          A general increase in air
          temperatures in our region of
          the world. And associated with
          that . . .

BACK TO DR. HINCH

                    DR. HINCH
          . . . then, would be a general
          increase in water temperatures.
          On top of that we also have . . .

CLOSE ON DR. HINCH

**LOWER THIRD: Dr. Scott Hinch, Professor,
Pacific Salmon Ecology, University of British
Columbia.**

                    DR. HINCH
          . . . oceanographic atmospheric
          issues that are going on at
          the same time. The two notable
          ones are the Pacific decadal
          oscillation, which is a, a
          phenomenon that persists for
          10 to 20 years at a time,
          switching between what we call
          regimes of high productivity
          and low productivity in the
          ocean, and associated with
          that are changes in coastal
          temperatures . . .

EXT. VEDDER RIVER MOUTH — DAY

Clear water mixes with murky water.

> DR. HINCH (O.S.)
> . . . going from either warm to
> cool, depending on which state
> you are in.

WIDER SHOT — showing two colours of water.

> DR. HINCH (O.S.)
> And what it's meant in the
> context of the Fraser over the
> last 20 years, how these all
> play together, is . . .

ON DR. HINCH

> DR. HINCH
> . . . a warming of the Fraser
> River and a warming of the
> coastal waters . . .

EXT. FRASER RIVER — DAY

The muddy river water mixes with clear water
from the Vedder River on a hot day.

> DR. HINCH (O.S.)
> . . . in southern British
> Columbia.

                    PATRICK MCGOWAN (O.S.)
               Explain to the Commissioner
               what the trends in the Fraser
               River have been in terms of
               temperature?

                    DR. HINCH (O.S.)
               Sure.

GRAPHIC — From page 90 of the report, showing
two lines, one red and one blue.

                    DR. HINCH (O.S.)
               This figure shows two lines.
               It's a relationship between
               average daily temperature in
               the lower Fraser River . . .

The rest of the graph appears.

                    DR. HINCH (O.S.)
               . . . and day, from the
               beginning of June to the end of
               September.

INSERT DATES — The dates "1951-1990" fade up
and indicate they belong to the blue line.

                    DR. HINCH (O.S.)
               And what it's showing is that
               from the early 1950s to 1990,
               the blue line, you can see what
               the average . . .

INSERT DATES — The dates "1991-2010" appear
and indicate they are associated with the red
line.

> DR. HINCH (O.S.)
> . . . daily temperature was.
> Um, since that period, in the
> recent period, from the early
> '90s to the present,

BACK TO GRAPH — Now the words "1990 to
present" fade up with an arrow pointing to
the red line.

> DR. HINCH (O.S.)
> . . . we have had, on average,
> about a degree warming, just
> under a degree warming
> throughout that entire time
> period.

BACK TO DR. HINCH

> DR. HINCH
> What is not shown on that
> figure in terms of the warming,
> because these are averages, is
> the extremes that we're now
> seeing, and we have many more
> extreme warm days in the past
> 20 years. In fact, 13 of the
> past 20 years have been the
> warmest on record.

PAN TO MR. MCGOWAN

                    PATRICK MCGOWAN
          Now, in terms of keeping in
          mind the warming temperatures,
          I'm wondering if you can
          briefly address for the
          Commissioner the significance
          of temperature to sockeye
          salmon?

                    DR. HINCH (O.S.)
          Okay.

ON DR. HINCH

                    DR. HINCH
          It controls everything from
          metabolism to physiology to
          behaviour to feeding, and
          there's really well-known
          relationships for many species
          about how temperature affects
          those processes. In sockeye
          salmon, in particular, with the
          adults that I'm focusing on
          with this figure, you can think
          of, of mortality and . . .

ON MR. MCGOWAN

                    DR. HINCH (O.S.)
          . . . survivorship as being
          related to two general
          processes.

BACK TO DR. HINCH

                    DR. HINCH
Things that kill you quickly,
or acute, and things that will
kill you slowly, or chronic.
The acute processes involved in
mortality usually are related
to how your metabolism or your
heart performance ceases. And
those things happen quickly at
certain temperatures. The more
chronically related effects
have to do with, um, diseases
and energy exhaustion, which
will take some time to take its
toll on individuals, depending
on what the water temperature
is.

                                    DISSOLVE TO:

INT. COURTROOM — AFTERNOON

Mr. McGowan is at his podium.

                    MR. MCGOWAN
Yes, Mr. Commissioner, I've
completed my examination. The
examinations this afternoon
will proceed in the usual order
with one exception and that is
Mr. McDade for the Aquaculture
Coalition is going first and
all other counsel who are
affected by that are agreeable.

CLOSE ON MR. MCDADE

> MR. MCDADE
> Starting in about 1992 is when
> you note this abrupt change in
> en route loss behaviour?

> DR. HINCH (O.S.)
> Um, starting in — yes, starting
> in 1992, en route loss really
> starts being reported, um . . .

TIGHT ON COMMISSIONER COHEN

> DR. HINCH (O.S.)
> . . . by the management
> agencies.

PULLING BACK FROM THE COMMISSIONER

> DR. HINCH (O.S.)
> By, in 1996 we start seeing a
> real large . . .

ON DR. HINCH

> DR. HINCH
> . . . or an abrupt change in
> the late-run sockeye en route
> loss values whereas prior to
> '96 it was minimal and after
> then it was, it was very large,
> owing to the early migration
> phenomenon. Prior to '92 en
> (MORE)

                    DR. HINCH (CONT'D)
          route loss wasn't really
          recorded or reported much,
          although it likely occurred in
          some years.

ON THE COMMISSIONER — He is making notes.

                    MR. MCDADE (O.S.)
          So it may have been occurring,
          but it was occurring in much
          smaller numbers?

                    DR. HINCH (O.S.)
          Correct.

TIGHT ON MR. MCDADE

                    MR. MCDADE
          I'm just still trying to get
          to a sense of the magnitude
          of these issues. If I come to
          paragraph 50, you've said . . .

EXHIBIT — On page 50. "Effects of en route
and pre-spawn mortality on population
trends."

                    MR. MCDADE (O.S.)
          . . . that the spawning
          abundance in early Stuart and
          Late run stocks, during a time
          period when . . .

MOVING IN ON THE DOCUMENT — The words
"en route loss had become a significant
component" are typed as they are highlighted.

                    MR. MCDADE (O.S.)
          . . . en route loss has become
          a significant component of the
          total fate —

                    DR. HINCH (O.S.)
          — Right.

                    MR. MCDADE (O.S.)
          It looks to me . . .

BACK TO MR. MCDADE

                    MR. MCDADE
          . . . that we're seeing in some
          years 50, 60, 70 percent —

                    DR. HINCH (O.S.)
          — Yes. —

                    MR. MCDADE
          — Loss?

                    DR. HINCH (O.S.)
          Yes.

Moments later . . .

                    MR. MCDADE
          Now, I just want to ask you one
          factual question.

GRAPH — of the en route loss by year.

>   MR. MCDADE (O.S.)
>   As I understand these charts,
>   that's en route loss?

>   DR. HINCH (O.S.)
>   That's correct.

ON MR. MCDADE

>   MR. MCDADE
>   Now, you also spoke about pre-
>   spawn mortality for those,

ON DR. HINCH

>   MR. MCDADE (O.S.)
>   . . . those fish, ah, that made
>   it to the spawning grounds and
>   then didn't spawn.

>   DR. HINCH
>   Right.

ON MR. MCDADE

>   MR. MCDADE
>   That would be additive to these
>   black lines, wouldn't it?

>   DR. HINCH (O.S.)
>   That's correct.

BACK TO THE GRAPH — The grey bars at the top turn black.

> MR. MCDADE (O.S.)
> It would be some proportion of the grey lines at the top of that chart?

> DR. HINCH (O.S.)
> Yes.

The two parts, including en route loss and pre-spawn mortality now in black, form a new graph that adds these two together.

> MR. MCDADE (O.S.)
> So if we were to combine these two numbers, en route loss and pre-spawn mortality, we're in numbers that exceed 70 percent?

> DR. HINCH (O.S.)
> Yes.

BACK ON MR. MCDADE

> MR. MCDADE
> And that would make this problem the single greatest problem in terms of loss of salmon of any that you're aware of, I would suggest.

ON DR. HINCH

                    DR. HINCH
          Yes it's, it's quite
          significant, quite significant
          level of non-spawning.

ON MR. MCDADE

                    MR. MCDADE
          If we go to, say, 2006, that in
          absolute numbers it can be as
          much as two million fish?

BACK TO DR. HINCH

                    DR. HINCH
          Mm-hmm. Yes.

BACK TO MR. MCDADE

                    MR. MCDADE
          So we could be looking at
          losses of over three million
          fish in some years?

BACK TO DR. HINCH

                    DR. HINCH
          Yes.

                    MR. MCDADE (O.S.)
          The term "critical," as I
          understand it,

ON MR. MCDADE

              MR. MCDADE
. . . is a fairly significant
one in science. What do you
mean by that?

TIGHTER ON MR. MCDADE

              DR. HINCH (O.S.)
Very important.

              MR. MCDADE
This might be the single
greatest causative factor we
have to look at?

ON DR. HINCH

              DR. HINCH
Yes.

ON MR. MCDADE

              MR. MCDADE
The, um, getting to the root
causes of early migration is a
fairly important question for
this Commission?

              DR. HINCH (O.S.)
It's one of them, yes.

                    MR. MCDADE
Because as I understand the
effect of temperature, you're
saying that the, the effect of
early entry in high-temperature
years can lead to increased
mortality?

                    DR. HINCH (O.S.)
Right.

                    MR. MCDADE
But it's, it's not the
temperature that causes the
early entry. It's the fact
of early entry into a high-
temperature environment?

                    DR. HINCH (O.S.)
Yes. Yes.

                    MR. MCDADE
Early entry you refer to at,
ah, at page 37, is an abrupt
shift in migration behaviour.

                    DR. HINCH (O.S.)
Mm-hmm.

                    MR. MCDADE
Now abrupt means sudden or
unexplained or —

                    DR. HINCH (O.S.)
          — Yes. It hadn't happened prior
          to '96 and suddenly this is
          occurring in large segments of
          the late runs.

                    MR. MCDADE
          All right. Um, and so we
          know that for the 60 years or
          more that we've been studying
          sockeye salmon in the Fraser
          River, that hasn't been
          happening and all of a sudden
          it starts?

                    DR. HINCH (O.S.)
          That's correct.

                    MR. MCDADE
          And climate change has
          been . . .

BACK TO DR. HINCH

                    MR. MCDADE (O.S.)
          . . . a steady and consistent
          matter moving throughout that
          60 years?

ON MR. MCDADE — He puts on his glasses.

                    DR. HINCH (O.S.)
          Whether it's been consistent, I
          don't know if I could agree to
          that.

COMMISSIONER COHEN — is making notes.

                MR. MCDADE (O.S.)
      Yes, but prior to 1992 there
      were warm years on record.

                DR. HINCH (O.S.)
      Yes.

PANNING TO MR. MCDADE

                MR. MCDADE
      And there was nothing in the
      climate change field that was
      sudden and abrupt in 1992?

                DR. HINCH (O.S.)
      No, no, not that I'm aware of.

                MR. MCDADE
      So we've got to look for some
      other causative factor?

                DR. HINCH (O.S.)
      Mm-hmm.

                MR. MCDADE
      So when we are looking
      at losses in the 50 to 70
      percent range, um, we have to
      recognize that for actually,
      if you segregate out the early
      entrants, you could see losses
      much higher than that, perhaps
      in the 90 percent range.

                    DR. HINCH (O.S.)
          Yes, in the context of the
          whole run,

EXT. HARRISON RIVER — DAY

Dead sockeye float in the river.

                    DR. HINCH (O.S.)
          . . . the earliest runs are
          the ones that are suffering
          the highest rates of
          mortality . . .

A dead sockeye floats near the shore.

                    DR. HINCH (O.S.)
          . . . and the more normal timed
          you become,

BACK TO DR. HINCH

                    DR. HINCH
          . . . the less the mortality
          rates would be on those fish.
          Right.

ON MR. MCDADE

                    MR. MCDADE
          I'd just like to go to the last
          page of that document.

DOCUMENT — slides into frame.

                    DR. HINCH (O.S.)
          Yep.

                    MR. MCDADE (O.S.)
          "Our hypothesis is that . . .

EXHIBIT — The words "Our hypothesis is that
the genomic signal associated with elevated
mortality is in response to a virus . . ."
pop up.

                    MR. MCDADE (O.S.)
          . . . the genomic signal
          associated with elevated
          mortality is in response to
          a virus infecting fish before
          river entry and that persists
          to the spawning areas."

BACK ON MR. MCDADE

                    MR. MCDADE
          You agree with that statement?

                    DR. HINCH (O.S.)
          Yes. The hypothesis is that.

PAN TO DR. HINCH

                    MR. MCDADE (O.S.)
          Yes. And, um, now, I'm
          wondering . . .

TIGHT ON MR. MCDADE

                    MR. MCDADE
          . . . why in the paper that you
          produced for this Commission
          the word "virus" does not
          appear.

                    DR. HINCH (O.S.)
          Right. Um. Two reasons. Ah, the
          first is that, um,

PAN OVER TO DR. HINCH

                    DR. HINCH
          . . . when I was writing the
          bulk of the paper, I was
          under a publication embargo.
          So I wasn't supposed to talk
          about or write about the
          Science paper. This is a, um,
          requirement of that particular
          journal.

FRONT PAGE — "Genomic Signatures Predict
Migration and Spawning Failure in Wild
Canadian Salmon."

                    DR. HINCH (O.S.)
          I nonetheless inserted the
          reference in so that it would
          get into the document so that
          we . . .

ON DR. HINCH

                    DR. HINCH
     . . . could talk about it.
     Um, the hypoth — as the
     paper suggests and, and as
     it's sprinkled throughout
     the *Science* paper, this is a
     hypothesis, and so I wanted
     to be clear in my report that
     I wrote that what we know for
     certain,

ON MR. MCDADE

                    DR. HINCH (O.S.)
     . . . absolutely certain, is
     that we're looking at an immune
     suppression response in, in the
     biochemical, the genomic data.

The Commissioner makes a note.

                    DR. HINCH (O.S.)
     That is a certainty.

                    MR. MCDADE (O.S.)
     So one reason for not referring
     to this directly in the
     paper . . .

TIGHT ON MR. MCDADE

                    MR. MCDADE
     . . . is the embargo that was
     Science just because the matter
     of timing?

                    DR. HINCH (O.S.)
Yes, awkward timing.

                    MR. MCDADE
Right. Did you discuss with
Commission counsel amending
your report to ah, to include
this?

                    DR. HINCH (O.S.)
Recently I did, very recently.

                    MR. MCDADE
And you were told it was too
late?

                    DR. HINCH (O.S.)
I was told, yeah, it was too
late.

                    MR. MCDADE
Otherwise you would have
included it?

                    DR. HINCH (O.S.)
If, if this would have taken
another month to bring
together, yes.

                    MR. MCDADE
So, ideally, you'd like to
amend your report to include
the possibility of this virus
as a causative factor?

> DR. HINCH (O.S.)
> I'm happy to talk about it
> right now.

> MR. MCDADE
> (laughs)
> Well, let's do that.

ON DR. HINCH — He smiles.

BACK TO MR. MCDADE

> MR. MCDADE
> Well, in terms of what is
> the causation of this early
> entry, this would be a fairly
> significant finding?

> DR. HINCH (O.S.)
> Yes.

> MR. MCDADE
> Yes. And the document goes on
> to say that in ocean tagged
> fish . . .

ON DR. HINCH

> MR. MCDADE (O.S.)
> . . . a mortality-related
> genomic signature was
> associated with a thirteen-and-
> a-half-fold greater chance of
> dying en route.

                    DR. HINCH
          Yes.

ON COMMISSIONER COHEN — He is making a note.

                    MR. MCDADE (O.S.)
          That's a very high number,
          isn't it?

                    DR. HINCH (O.S.)
          Very high number.

PUSHING IN — on the Commissioner's screen

                    MR. MCDADE (O.S.)
          Now, when Mr. Commissioner asked
          you this morning a question
          about how can we tell which
          fish are going to die and which
          ones aren't — I'm sorry to
          paraphrase, Mr. Commissioner —

BACK ON COMMISSIONER COHEN

                    MR. MCDADE (O.S.)
          . . . this is a pretty
          significant answer.

                    DR. HINCH (O.S.)
          Yeah. And that's — I knew you
          were going to ask me this, so
          I was leading you into this by
          saying that there are
                    (MORE)

                    DR. HINCH (CONT'D)
          physiological conditions that
          can predispose an animal to its
          fate. In this case,

ON MR. MCDADE

                    DR. HINCH (O.S.)
          . . . um, there was a suite
          of genes that represented a
          particular . . .

BACK TO DR. HINCH

                    DR. HINCH
          . . . physiological state that
          was predictive of what it was
          going to do later in its life.
          In this case, perish.

PANNING TO MR. MCDADE

                    MR. MCDADE
          So this purported virus, if it
          in fact exists —

                    DR. HINCH (O.S.)
          — Mm-hmm —

                    MR. MCDADE
          — goes a very substantial way
          towards explaining the early,
          or to explain the whole of the
          en route loss?

PANNING BACK TO DR. HINCH

> DR. HINCH
> It could. And that's why it
> got published in the journal
> *Science*, because they're
> looking for these broad-scale
> wow sorts of relationships.

FADE OUT:

CHAPTER 6

The Question of Taking Action.

Or,

"How much proof is needed

before we do something?"

Two bloated Harrison River sockeye lie on the beach at Kilby Park (September 10, 2011). Both fish died before being able to spawn, a phenomenon known as pre-spawn mortality. Mysterious mass die-offs have plagued BC's Pacific salmon populations for over 40 years, and the Harrison River was one of the last rivers to experience a catastrophic die-off. (Photo: Scott Renyard)

INT. COURTROOM — DAY

SUPERSCRIPT: August 25, 2011 . . .

Dr. Garver is on the stand.

> MR. MCDADE (O.S.)
> Dr. Garver, I just wanted to
> draw your attention.

ON MR. MCDADE

> MR. MCDADE
> I'm sure you won't disagree,
> since you said it, but lab
> studies indicate that farmed
> fish can shed, um, up to
> 200,000 viral particles per
> fish per hour. That's accurate?

PAN TO DR. GARVER

> DR. GARVER
> This was a study that I've been
> currently conducting on the IHN
> virus, and we're interested in
> the transmission and dispersion
> of this virus.

ON MR. MCDADE

                    MR. MCDADE
          So in terms of real numbers, Dr.
          Garver, here is a presentation
          you gave showing, I guess, in
          theory that at a maximum level a
          fish farm could be shedding up
          to 60 billion viral particles
          per hour.

PAN BACK TO DR. GARVER

                    DR. GARVER
          Yeah, this is what we call
          a back-of-the-envelope
          calculation here. Again, this
          was in a laboratory study,

INSERT — Exhibit 1529 and the back-of-the-
envelope calculation.

                    DR. GARVER (O.S.)
          . . . but that's based on . . .

WHIP TO DR. GARVER

                    DR. GARVER
          . . . a peak shedding rate. So
          when you have maximum die-off,

ON THE GALLERY — It's full.

                    DR. GARVER (O.S.)
          . . . right before that die-
          off . . .

BACK TO DR. GARVER

>   DR. GARVER
>   . . . your, the fish would be
>   shedding the most amount of
>   virus.

ON MR. MCDADE

>   DR. GARVER (O.S.)
>   If you have . . .

BACK TO DOCUMENT

>   DR. GARVER (O.S.)
>   . . . a farm that has
>   approximately a million
>   fish . . .

The formula is highlighted.

>   DR. GARVER (O.S.)
>   . . . and they're
>   experiencing . . .

TO DR. GARVER

>   DR. GARVER
>   . . . a 30 percent infection,
>   which based . . .

MR. MCDADE — checks the monitor.

> DR. GARVER (O.S.)
> . . . on some of the die-off
> events . . .

BACK TO DR. GARVER

> DR. GARVER
> . . . and then you times that by
> the number of particles . . .

BACK TO FORMULA — Highlight "60 billion viral
particles shed per hour."

> DR. GARVER (O.S.)
> . . . that we quantified in
> the water, and you do get 60
> billion viral particles shed
> per hour.

ON DR. MILLER

> MR. MCDADE (O.S.)
> Dr. Garver, Dr. Miller strikes
> me as a relatively level-headed
> person . . .

ON THE CROWD — listening intently.

> MR. MCDADE (O.S.)
> . . . not prone to, ah, Chicken
> Little, the-sky-is-falling
> material.

ON DR. MILLER — also listening.

>   MR. MCDADE (O.S.)
> When a senior scientist at your
> department says . . .

BACK ON MR. MCDADE

>   MR. MCDADE
> . . . "potentially devastating
> impacts," um, that's a
> significant finding for you, is
> it not?

TO DR. GARVER

>   DR. GARVER
> I'm sorry, for me? You're,
> you're, um . . .

>   MR. MCDADE (O.S.)
> Well, what I'm trying to do is
> get to the sense of what level
> of certainty do you need . . .

PANNING BACK TO MR. MCDADE

>   MR. MCDADE
> . . . about a potentially
> devastating impact to the
> sockeye salmon to actually take
> action . . .

ON DR. GARVER

                    MR. MCDADE (O.S.)
          . . . rather than more studies?
          What would it take to get you
          to actually recommend some
          action?

BACK ON MR. MCDADE

                    MR. MCDADE
          How far do we have to go in
          proof?

PAN BACK TO DR. GARVER

                    DR. GARVER
          We're following a scientific
          approach. So we need to
          establish that this sequence is
          indeed causing a disease.

PAN BACK TO MR. MCDADE

                    MR. MCDADE
          And you're not prepared to
          recommend action to the senior
          people at DFO until you've done
          all of those laboratory studies
          and have found proof to your
          satisfaction?

PAN OVER TO DR. GARVER

                    DR. GARVER
     Until I find that this virus is
     causing disease and that it is
     associated with the MRS,

POP-UP — MRS: Mortality Related Signature

                    DR. GARVER
     . . . and that it is indeed
     transmissible, then I probably
     would not recommend action at
     this time.

ON THE CROWD

                    MR. MCDADE (O.S.)
     Well, Dr. Miller was, was at
     least hypothesizing . . .

BACK TO MCDADE

                    MR. MCDADE
     . . . that some 27 million
     salmon might have died from
     this in 2008. Wouldn't that be
     something that you would take
     some action about?

BACK TO DR. GARVER

                    DR. GARVER
     And that is indeed what we're
     doing. We're researching
     whether this sequence causes
     disease.

BACK ON MR. MCDADE

> MR. MCDADE
> So for you, action, when
> millions of salmon are dying,
> is to take research.

TO MR. TAYLOR

> MR. TAYLOR
> Well, I'm going to get up
> at this point. Mr. McDade
> is asking a question of a
> scientist and the scientist is
> answering. Mr. McDade wants an
> answer from a manager, but a
> manager's not on the stand.

BACK TO MR. MCDADE

> MR. MCDADE
> So this is not, do you feel,
> do you agree with that, Dr.
> Garver, that this is not your
> business as to what action is
> taken?

ON DR. GARVER

> DR. GARVER
> The management is aware of
> these briefing notes, these
> memos. I conduct science.

                    MR. MCDADE (O.S.)
          Well, we saw yesterday in a
          memo that Dr. Miller prepared
          to go to management, we saw all
          those comments from you trying
          to water that down. Why would
          you try and resist her telling
          senior management what her
          views were?

                    DR. GARVER
          I gave my scientific opinion.
          That's what I, that's my job. I
          weigh the evidence. And I put
          it out there. That's what they
          hire me for. I am a scientific
          person.

PAN TO MR. MCDADE

                    MR. MCDADE
          Well, when, in the public
          health field, when SARS was
          first discovered to be killing
          human beings at some risk, the
          virus hadn't been cultured and
          proven to the levels that you
          talk about, had it?

EXT. HARRISON RIVER — DAY

Dead sockeye float in the water.

                    DR. GARVER (O.S.)
          It had been identified.

                    MR. MCDADE (O.S.)
          But —

EXT. SHORELINE — DAY

Two silver sockeye lie dead on the shore.

                    DR. GARVER (O.S.)
          — And it was definitively
          linked to the disease.

ON DR. GARVER

                    MR. MCDADE (O.S.)
          My suggestion to you is public
          health officials, when the
          health of human beings is
          involved, take action before the
          final proof of the virus is in.

PAN TO MR. TAYLOR

                    MR. TAYLOR
          This witness hasn't been put up
          as a public health expert, and
          Mr. McDade hasn't done anything
          to establish that he's going to
          have any basis of knowing what
          the question's about.

OVER TO MR. MCDADE

                    MR. MCDADE
          Well, let me re-ask the
          question, then, in a different
          way.

TO DR. GARVER

                    MR. MCDADE (O.S.)
          Do you have any guide in your
          department that suggests that
          you should take action in the
          absence of final proof? What
          level of risk does it take to
          actually start doing something?

                    DR. GARVER
          I believe we are doing quite a
          bit.

OVER TO DR. MILLER

                    MR. MCDADE (O.S.)
          Let me ask you, then, Dr.
          Miller. Why didn't you go out
          and test in 2009?

DR. MILLER — answers.

                    DR. MILLER
          The feeling was that we didn't
          have an etiological agent. We
          hadn't identified an actual
          pathogen at that time that I
          was discussing this in 2008
                    (MORE)

                    DR. MILLER (CONT'D)
          and 2009. And I think it, it
          was very difficult, um, to
          get across to the fish health
          community, you know, what is a
          genomic signature, what is the
          power of a genomic signature,
          how much can we derive in terms
          of, of realistic mechanisms
          from a genomic signature. And
          so the, the, the battle really
          was, you know, um, until you
          have an etiological agent we
          really can't, you know, ask
          industry to test. We really
          can't move forward.

PAN TO MR. MCDADE

                    MR. MCDADE
          Well, I may have misunderstood
          your testimony yesterday, but
          I heard that you tried to get
          the fish farmers to let you
          test and their veterinarians
          refused.

PAN BACK TO DR. MILLER

                    DR. MILLER
          I was approached by Mary Ellen
          Walling right . . .

Camera settles on Dr. Miller.

                    DR. MILLER
. . . after the *Science* paper
came out, probably in early
February, and she wanted to
know more about what we knew
and what we had, and she said
there was some interest in the
industry to go ahead and test
for the signature we have. Um,
I was told later by one of the
vets, by one of the companies,
that they were advised against
doing the testing by someone
from DFO. So that's as far as
it went.

ON MR. MCDADE

                DR. MILLER (O.S.)
Um, and I approached . . .

BACK TO DR. MILLER

                    DR. MILLER
. . . the Fish Health group in
Laura Richard's office again
in, at the end of July, about
moving forward and testing.
And, and really my question was
who would ask the industry, ah,
to provide samples for testing.
Would that come from me, would
that come from management,
would that come from Fish
                    (MORE)

                    DR. MILLER (CONT'D)
          Health, and how do we move
          forward with this? And we
          had a discussion in Laura
          Richard's office about that.
          And, um, it was clear from that
          discussion . . .

MONTAGE — Emails.

                    DR. MILLER (O.S.)
          . . . that it was a decision
          to be made by the Fish Health
          group. And at the time, they
          were still uncomfortable
          with, with asking the industry
          to test, and that's what
          those . . .

DOCUMENT — with a box that is a summary of
some kind.

                    DR. MILLER (O.S.)
          . . . emails, subsequent emails
          were about.

DOCUMENT — Tight on the phrase "DFO has
requested more extensive sea lice and fish
disease data from the fish farm operating in
all areas of the BC coast."

                    MR. MCDADE (O.S.)
          So, the question I had, though,
          is you were briefing Dr. Laura
          Richards, weren't you?

                    DR. MILLER (O.S.)
          I don't recall if Dr. Laura
          Richards was there.

Mr. McDade steps away from his podium in
disgust.

                    DR. MILLER (O.S.)
          Certainly Mark Saunders, the
          head of SAFE Division was
          there. He's the one who has,
          has taken sort of the lead on,
          on this PARR program. I don't
          recall if Laura Richards was
          there or not. If she was on the
          list she was there, but I don't
          recall.

                    MR. MCDADE
          So, why would you be briefing
          the salmon farmers? You can't —

PAN OVER TO DR. MILLER

                    DR. MILLER
          — Again, I didn't know that
          the salmon farmers were there.
          This, you know, they have a,
          what I would say, a, a role in
          this program, but I was unaware
          if the salmon farmers were
          there or not.

BACK ON MR. MCDADE

                    MR. MCDADE
          So when Dr. Richards testified
          here in March, in March,
          though, she had just been
          briefed with you, from you
          about the parvovirus?

TO MR. TAYLOR

                    MR. TAYLOR
          Well, that's not the evidence.
          The evidence is Dr. Miller
          doesn't know if Dr. Richards
          was there.

TO DR. MILLER

                    DR. MILLER
          I know, can I, I did talk to
          Dr. Richards on that day. I
          don't know that she was at
          that particular meeting, but
          that meeting came very shortly
          after we had really truly,
          um, discovered this, this
          parvovirus, um, and I briefed
          Mark Saunders in the morning
          about that. We had this meeting
          in the afternoon, and I believe
          I had a second meeting in Laura
          Richards's office with Mark
          Saunders, ah, later in that
          afternoon. So what I don't know
          is if Laura Richards was at
                    (MORE)

                    DR. MILLER (CONT'D)
the meeting for the PARR and I
honestly did not know at the
time that there was anybody
associated with the aquaculture
industry in that room. And in
fact what I was told from Andy
Thomson when he had talked to
the, the leads of the various
companies that the only company
that had an inkling that we had
a virus was Marine Harvest.

PAN BACK TO MR. MCDADE

                    MR. MCDADE
Well, I have much more on
this, Mr. Commissioner, but
regretfully I'm out of time.

                              FADE OUT:

CHAPTER 7

The Question of Examining

All Possibilities.

Or,

"Was the review about

finding an excuse?"

From:	Richards, Laura
Sent:	Tuesday, October 6, 2009 12:42 PM
To:	Saunders, Mark <Mark.Saunders@dfo-mpo.gc.ca>
Subject:	RE: Draft BN
Attach:	Sockeye2009_bn_min_e_MS1 (5).doc

<<...>>

Dr. Laura Richards
Regional Director Science | Directrice régionale des sciences
Fisheries and Oceans Canada | Pêches et Océans Canada
Pacific Biological Station | Station biologique du Pacifique
3190 Hammond Bay Rd, Nanaimo, BC, Canada V9T 6N7

Laura.Richards@dfo-mpo.gc.ca
Telephone | Téléphone 250-756-7177
Facsimile | Télécopieur 250-729-8360
Government of Canada | Gouvernement du Canada

From: Saunders, Mark
Sent: October 6, 2009 10:50 AM
To: Richards, Laura
Subject: Draft BN

Hi Laura,

Still needs summary bullets. Hopefully this is closer. Could easily move the hypotheses to an attachment if need be. <<
File: Sockeye2009_bn_min_e_MS1 (5).doc >>

Mark

Mark W. Saunders
Division Manager, Salmon and Freshwater Ecosystems
Directeur de secteur, Saumons et Ecosystèmes d'eau Douce
Fisheries and Oceans Canada / Pêches et Océans Canada
Pacific Biological Station / Station Biologique du Pacifique
3190 Hammond Bay Rd, Nanaimo, BC V9T 6N7

tel/tél: 250-756-7145
FAX: 250-756-7053
Mobile/tél cell: 250-713-3786
e-mail/courriel: Mark.Saunders@dfo-mpo.gc.ca

The Department of Fisheries and Oceans staff examined a number of possible factors—including unusual winds—that could have caused the decline of Fraser River sockeye. An email presented at the Inquiry revealed that the department was looking for something to counter Alexandra Morton's data. It was not clear from the email whether the department was trying to be thorough or was looking for another reason for sockeye salmon declines in order to divert attention away from open net pen aquaculture. (Source: Cohen Commission, Exhibit 615)

INT. COURTROOM — DAY

Commissioner Cohen is reading a document.

SUPERSCRIPT: March 17, 2011 . . .

ON DR. RICHARDS — on the stand.

             MR. MCDADE (O.S.)
     Can I go to page 9? You see the
     reference . . .

DOCUMENT — Exhibit 613G, page 9, "Tumor-Associated in-River Mortality of Adult Sockeye Salmon."

PUSH IN — "2008 unprecedented levels of mortality."

             MR. MCDADE (O.S.)
     . . . to unprecedented levels
     of mortality in 2008?

ON MR. MCDADE — He looks over at Dr. Richards.

             MR. MCDADE
     Do you see that?

             DR. RICHARDS (O.S.)
     Yes.

                    MR. MCDADE
     And the reference below that is
     to the 2009 sockeye not showing
     up, which is the purpose of
     this Commission, and, um, the
     suggestion that . . .

ON DR. RICHARDS

                    MR. MCDADE (O.S.)
     . . . the 20 percent decline in
     tumours could account for point
     nine million fish going missing
     in the . . .

BACK ON MR. MCDADE

                    MR. MCDADE
     . . . Strait of Georgia, right?

                    DR. RICHARDS (O.S.)
     That's what that says, although
     I think we have to be a . . .

PAN TO DR. RICHARDS

                    DR. RICHARDS
     . . . little careful in doing
     an interpretation of, of a,
     this kind of presentation
     because we don't have all the
     full context and the facts
     stated here.

                    MR. MCDADE (O.S.)
          True. But the, in terms of what
          DFO science knew, this was the
          best information you had in
          September 2009?

                    DR. RICHARDS
          Well, this was, this was
          the, her interpretation at
          that time. I mean, this is
          very much, has been a work in
          progress, and I think we need
          to be really clear that the
          work that she's been doing
          here has really been on, on the
          forefront and when you're on
          the forefront, you don't always
          have all, it isn't fully worked
          out and this particular area
          is something, this area has
          been one which has been quite
          evolving in the thinking.

PAN BACK TO MR. MCDADE

                    MR. MCDADE
          Yes. I'm just going to what, in
          the evolution, September 30th,
          2009, this was the thinking
          of your top scientist on this
          matter?

                    DR. RICHARDS (O.S.)
          That was her thinking at that
          time.

                    MR. MCDADE
          Right. And that she noted
          that in 2005, which was the
          brood year for the 2009, that
          75 percent of the adults
          were positive for that viral
          signature, right?

PAN BACK TO DR. RICHARDS

                    DR. RICHARDS
          That's what this — yes. You're
          reading the document, yes.

                    MR. MCDADE (O.S.)
          And —

                    DR. RICHARDS
          — That's what —

                    MR. MCDADE (O.S.)
          — Well, I'll go, go in a few
          minutes to what you did with
          this material, but, ah, I just
          wanted to know what material
          you had to work with.

DOCUMENT — Page 11, the words "Strong
Linkages of Genomic and Brain Tumour Data"
and then "with Plasmacytoid Leukemia
Caused by the Salmon Leukemia Virus" are
highlighted.

                    MR. MCDADE (O.S.)
          So if I could go to page 11.
          So the heading of that page
          is "Strong Linkages of Genomic
          and Brain Tumour Data with
          Plasmacytoid Leukemia Caused by
          the Salmon Leukemia Virus."

DISSOLVE TO MR. MCDADE

                    MR. MCDADE
          That, that was the hypothesis
          of your best scientist on this
          matter?

PAN BACK TO DR. RICHARDS

                    DR. RICHARDS
          That was her hypothesis. I'm
          not, I, as I said, there was
          some other points of view
          amongst some of our other
          experts in fish health on this.

ON MR. MCDADE

                    MR. MCDADE
          Well, did either of those
          people disagree with Dr.
          Miller?

BACK TO DR. RICHARDS

                    DR. RICHARDS
          I think what we want to do is
          just to try to be, you know,
          careful. Some of this was
          really on speculation and what
          we're really trying to do . . .

COMMISSIONER COHEN — looks up from his notes.

                    DR. RICHARDS (O.S.)
          . . . is ground this in fact.

MOVING OVER TO DR. RICHARDS

                    MR. MCDADE (O.S.)
          But the name of the virus that
          was being purported at that
          point . . .

BACK TO MR. MCDADE

                    MR. MCDADE
          . . . was the salmon leukemia
          virus, right?

                    DR. RICHARDS (O.S.)
          That was certainly, was a
          hypothesis that she was putting
          forward.

                    MR. MCDADE
          This is also called — you see
          the third bullet, it's also
          called marine anemia?

                    DR. RICHARDS (O.S.)
Yes.

                    MR. MCDADE
That's the name for it when
it's in fish farms, right?

                    DR. RICHARDS (O.S.)
I am not the expert on diseases,
um, so I'm sure we'll,

PAN BACK TO DR. RICHARDS

                    DR. RICHARDS
. . . you'll have more
testimony on that later, I
would expect.

                    MR. MCDADE (O.S.)
Well, when you were preparing
the briefing note, did you know
that?

                    DR. RICHARDS
Well, I, I think what I was
doing was relying on those
experts to help prepare that
note, and I was clarifying
them that we did have that
technically correct.

BACK TO MR. MCDADE

                    MR. MCDADE
          The heading on page 14 . . .

                                        DISSOLVE TO:

DOCUMENT — On the words "If SLV is a primary
factor in the salmon declines in BC."

                    MR. MCDADE (O.S.)
          . . . refers to "If SLV" —
          that's salmon leukemia virus
          — "is a primary factor in the
          salmon declines in BC."

BACK TO MR. MCDADE

                    MR. MCDADE
          There was at least speculation
          by Dr. Miller that the salmon
          leukemia virus was a primary
          factor in the salmon declines.

PAN TO DR. RICHARDS

                    DR. RICHARDS
          That was her speculation at the
          time as I recall.

                    MR. MCDADE (O.S.)
          Yes. Now, under, under the
          heading "Potential for
          mitigation," you'll see that
          she refers to the possibility
                    (MORE)

                    MR. MCDADE (CONT'D)
          that if one takes action, you
          could minimize the vertical
          transmission.

DOCUMENT — Exhibit 613G. The words "minimize
vertical transmission" are highlighted.

                    MR. MCDADE (O.S.)
          What do you as a scientist
          understand the words "vertical
          transmission" to mean in the
          salmon field?

BRIEFLY ON DR. RICHARDS

ON MR. MCDADE

                    DR. RICHARDS (O.S.)
          As I said, this is not my area
          and I just want to be very
          careful to not give incorrect
          evidence.

DOCUMENT 635 — November 13, 2008. Briefing
note. Cover page.

                    MR. MCDADE (O.S.)
          Over the page then, just
          underneath the table.

ON THE WORDS — "the potential devastating
impacts of this disease."

                    MR. MCDADE (O.S.)
          Dr. Miller here refers to it in
          this document as "the potential
          devastating impacts of this
          disease on sockeye salmon." You
          see those words?

                    DR. RICHARDS (O.S.)
          Umm.

                    MR. MCDADE (O.S.)
          Just below the table.

                    DR. RICHARDS (O.S.)
          Okay, yes.

ON MR. MCDADE

                    MR. MCDADE
          Yes. Now, I think I asked
          you on November 4th when you
          were last here, when one,
          when science is dealing with
          something that is a potentially
          devastating impact, do you act
          any differently? Do you, do you
          fund more research? Do you move
          quicker?

                    DR. RICHARDS (O.S.)
          I would hope that we would do
          that.

ON DR. RICHARDS

                    DR. RICHARDS
          We were extremely interested,
          and I don't, certainly were
          aware of these potential
          consequences but we also wanted
          to try to get more information
          and were wanting to work with
          Dr. Miller so that we could get
          more substantive information
          to understand this and also to
          understand the scope because,
          as I say, once you start to
          look at this, you try to, you
          certainly come up with more
          questions.

DOCUMENT — Next page.

                    MR. MCDADE (O.S.)
          If I could go over the page to
          the second paragraph on the
          next page, starting, "The very
          significant reduction."

PUSH IN — The words "may be indicative" are
highlighted.

                    MR. MCDADE (O.S.)
          It suggests the en route
          mortality may be indicative of
          lesion-associated mortality.

ON MR. MCDADE

                    MR. MCDADE
And that, "If so, the levels
of mortality required to bring
prevalence levels down . . .

BACK TO DOCUMENT — The words "large-scale
losses" are highlighted.

                  MR. MCDADE (O.S.)
. . . by over 30 percent would
be sufficient in magnitude to
account for large-scale losses
in the ocean."

ON DR. RICHARDS

                  MR. MCDADE (O.S.)
Do you see that?

                  DR. RICHARDS
Yes.

ON MR. MCDADE

                  MR. MCDADE
And if I could go over to
Exhibit 615, we'll look at
that.

DOCUMENT — Exhibit 615, email.

                  MR. MCDADE (O.S.)
So that's an email dated
October 6th with an
attached . . .

ON THE DOCUMENT — Push in and highlight the words "Factors affecting sockeye returns."

> MR. MCDADE (O.S.)
> . . . draft of the factors affecting sockeye . . .

BACK TO MR. MCDADE

> MR. MCDADE
> . . . return on the next page. I just want to take you through your conclusions at this time. Yours or Mr. Saunders's, whichever it was. The first, "Pollution in Lake Watersheds," that was said to be unlikely for, as the cause of the 2009 decline, right?

> DR. RICHARDS (O.S.)
> Yes.

> MR. MCDADE
> The second, "Fishery Effects," it says "possible impacts." But if you look at the last line,

TURN THE PAGE — Move down to the bullet. Push in and highlight the words "would only explain a small portion of the 2009 mortality."

PAN TO DR. RICHARDS

                    MR. MCDADE (O.S.)
          . . . the conclusion was it
          would only explain a small
          proportion of the 2009
          mortality. That was your
          conclusion at the time.

                    DR. RICHARDS
          Yes.

ON MR. MCDADE

                    MR. MCDADE
          The third bullet is the viral
          disease effects.

BACK TO THE DOCUMENT — Pan down to the viral
disease bullet and highlight the words "virus
that could be a major contributor."

                    MR. MCDADE (O.S.)
          And the conclusion there is
          the virus could be a major
          contributor to the mortality
          occurring through the life
          history.

BACK ON MR. MCDADE

                    MR. MCDADE
          And, um, that it could provide
          an explanation for the short-
          and long-term declines for
          sockeye.

ON DR. RICHARDS

           MR. MCDADE (O.S.)
  Is that right?

            DR. RICHARDS
  Yes.

MOVING DOWN THE DOCUMENT — "Predation
offshore by Humboldt squid."

           MR. MCDADE (O.S.)
  The next bullet is . . .

PHOTO — of Humboldt squid.

           MR. MCDADE (O.S.)
  . . . "Predation offshore by
  Humboldt squid."

BACK TO DOCUMENT — On the words "not a key
factor."

           MR. MCDADE (O.S.)
  It said that that's a possible
  impact. But down the paragraph
  it suggests that it's not a key
  factor.

           DR. RICHARDS (O.S.)
  Yes.

BACK TO MR. MCDADE

                    MR. MCDADE
The next one is "Predation in
Strait of Georgia." And your
conclusion then was that that
was unlikely to be the impact
for 2009, right?

                    DR. RICHARDS (O.S.)
Yes.

                    MR. MCDADE
And the next one was sea lice.
And what you say there is it's
possible. Let me paraphrase
the first sentence. While
it is possible that sea lice
from farms contributed to the
mortality, the degree of impact
is difficult to assess. So
you weren't attributing the
2009 cause to sea lice at that
point?

ON DR. RICHARDS

                    DR. RICHARDS
That's correct.

                    MR. MCDADE (O.S.)
The next heading on the
following page is the mortality
attributed to algal blooms.
So your conclusion again was
although it was a possible
impact, it was unlikely.

                    DR. RICHARDS
          Um, well, that's what is
          written here but that's not
          consistent with what I think we
          have a little bit later.

                    MR. MCDADE (O.S.)
          Right. But at this time, your
          conclusion —

ON MR. MCDADE

                    DR. RICHARDS (O.S.)
          — Well, this is, this is the
          work that he had just put in.
          That's what he has written in
          this note.

                    MR. MCDADE
          All right. The next phrase
          related to the krill fishery.
          And there again the reference
          is that the 2009 impact was
          unlikely.

ON DR. RICHARDS

QUICKLY BACK TO MR. MCDADE

                    DR. RICHARDS (O.S.)
          Yes.

                    MR. MCDADE
          And the mortality due to low
          prey abundance,

PAN TO DR. RICHARDS

> MR. MCDADE (O.S.)
> . . . there doesn't seem to be
> a rating in that paragraph, but
> that's the one that you said
> subsequently lessened, in your
> view, as being a significant
> cause.

ON MR. MCDADE

> DR. RICHARDS (O.S.)
> I think at this point we would
> rate . . .

BACK TO DR. RICHARDS

> DR. RICHARDS
> . . . the Strait of Georgia
> issues as probably higher than
> those, than the factors that
> happened in Queen Charlotte
> Sound.

PAN TO MR. MCDADE

> MR. MCDADE
> Now, I note under
> "Recommendations," the only
> specific one, of course, is,
> that's dealt with is exactly
> that, the one that's probably
> most likely, the disease work,
> the second bullet, right?

                    DR. RICHARDS (O.S.)
Yes.

                    MR. MCDADE
And what, what is said there
by you and Mr. Saunders is the
disease work will be of extreme
interest . . .

ON DR. RICHARDS

                    MR. MCDADE (O.S.)
. . . and may be quite
controversial.

ON MR. MCDADE

                    MR. MCDADE
Now, can you explain to the
Commissioner what you meant by
"extreme interest" and why?

PAN TO DR. RICHARDS

                    DR. RICHARDS
Well, okay, first of all, it's
not what I meant. As I said,
this was the first draft of
a note that was done by Mr.
Saunders. But I think what's
intended here is we do think
that — I mean, I do think we
think that the work that Kristi
                    (MORE)

                    DR. RICHARDS (CONT'D)
          Miller has done is, ah,
          very, um, important and —
          first of all, it's very
          interesting . . .

TIGHT ON MR. MCDADE

                    DR. RICHARDS (O.S.)
          . . . scientifically but also
          very important from trying to
          contribute an understanding to
          what's going on here.

ON DR. RICHARDS

                    MR. MCDADE (O.S.)
          And by "controversial," did
          you understand that to mean
          controversial in terms of the
          public reaction?

                    DR. RICHARDS
          Yes.

                    MR. MCDADE (O.S.)
          And the next line says a
          communication strategy has to
          be developed. Or should be
          developed. What did that, what
          did that mean, a communication
          strategy?

ON MR. MCDADE

                    DR. RICHARDS (O.S.)
          Well, I mean, this is a
          standard piece. When we have
          some new findings or new
          information,
               (she stammers for a moment)
          we would normally prepare some
          kind of briefing note.

BACK TO DR. RICHARDS

                    DR. RICHARDS
          We would normally help staff in
          preparing communication lines
          or media lines around this kind
          of information.

ON MR. MCDADE

                    MR. MCDADE
          Well, um, were you concerned
          about public over-reaction?

PAN TO DR. RICHARDS

                    DR. RICHARDS
          Well, we were concerned about
          trying to get as much of the
          truth out as possible. And we
          were concerned about trying to
          be as, as factual as we could
          and trying to figure out what
                    (MORE)

                    DR. RICHARDS (CONT'D)
really was going on because,
ah, yes, I think we were
concerned about over-reaction,
but —

BACK ON MR. MCDADE

                    MR. MCDADE
— Was a communication plan
prepared?

                    DR. RICHARDS (O.S.)
I don't recall that one was
prepared.

                    MR. MCDADE
Let me suggest to you that the
communications plan was not to
release this to the public at
all.

PAN TO DR. RICHARDS

                    DR. RICHARDS
Ah, that is false.

                    MR. MCDADE (O.S.)
When did the department first
release to the public that
there was a suspected virus in
salmon?

                    DR. RICHARDS
          Well, I think Kristi Miller has
          been going to various meetings
          and talking to that. And there
          was some discussion in various
          forums about this.

PANNING BACK TO MR. MCDADE

                    MR. MCDADE
          My question to you relates
          to releasing this data to the
          public or to the media, not to
          scientific forums.

                    DR. RICHARDS (O.S.)
          Well, I mean, our job is not to
          release things directly to the
          media. That's not the, that's
          not the role of science within
          the department. What we, what I
          do, do is try to, to inform the
          decision-makers.

ON COMMISSIONER COHEN

THEN BACK TO MR. MCDADE

                    MR. MCDADE
          The department doesn't write
          letters to the media relating
          to the sea lice threat, for
          instance?

PAN TO DR. RICHARDS

>   DR. RICHARDS
>   Yes, the department writes
>   letters. But I'm just, but —

TIGHT ON MR. MCDADE

>   MR. MCDADE (O.S.)
>   — Let me suggest to you the
>   first time this question of
>   the virus has ever been in the
>   public record in the media was
>   November 3rd, 2010, when the
>   *Globe and Mail* released this
>   memorandum.

PAN TO MR. TAYLOR

>   MR. TAYLOR
>   Well, I object to the question.
>   How is the witness supposed
>   to know when the media did
>   something? I doubt Mr. McDade
>   knows when the media did
>   everything that they do.

BACK ON MR. MCDADE

>   MR. MCDADE
>   Well, I'll rephrase the
>   question. Do you know whether
>   the media has ever been told of
>   this disease?

ON DR. RICHARDS'S AND MR. TAYLOR'S REACTIONS

THEN ON DR. RICHARDS

> DR. RICHARDS
> I think, Mr. Commissioner, I'm
> not really interested, or that
> my job has not got to do with
> what the media knows or doesn't
> know.

TIGHT ON MR. MCDADE

> DR. RICHARDS (O.S.)
> Our process would not be . . .

BACK TO DR. RICHARDS

> DR. RICHARDS
> . . . to go directly to the
> media on this. Our process, but
> would be to prepare scientific
> papers that ah, that follow
> through a normal peer-review
> process.

FADE OUT:

CHAPTER 8

The Question of the Press.

Or,

"Was the Conservative Government

muzzling its scientists?"

The Harper administration began to curtail the flow of scientific research in 2006. In 2007, the government set up new rules that "controlled Environment Canada scientists' interactions with the media" (Linnitt, May 2013). In 2013, it shut down the Experimental Lakes Area project in Ontario and then began shutting down 7 of the 11 Department of Fisheries and Oceans libraries in 2014 (Galloway, 2014). All of this behaviour was labelled by the press as the muzzling of Canadian scientists. (Photo: Scott Renyard)

PAN TO MR. MCDADE

          MR. MCDADE
Can I go to Exhibit 628,
please?

          MR. WALLACE (O.S.)
Mr. Commissioner, I just note
that by the allocation, Mr.
McDade has eight minutes.

          MR. MCDADE
Well, I count nine, but I'll
see what we can do.

Laughter erupts from the gallery.

          MR. MCDADE
Exhibit 628, please?

ON DR. RICHARDS

          MR. MCDADE (O.S.)
This is the document, Dr.
Richards, where . . .

ON EXHIBIT 628 — the email. The words "Laura
also clearly does not want to indicate to the
PSC that the disease research is of strategic
importance" are highlighted.

> MR. MCDADE (O.S.)
> . . . Dr. Miller is quoted as
> saying that you don't want to
> indicate to the PSC that
> disease research is of
> strategic importance. Where did
> she get that idea?

BACK TO DR. RICHARDS

> DR. RICHARDS
> I don't know where she got that
> impression and that impression
> that she has is completely
> false.

BACK TO EMAIL — "Laura does not want me to attend"

> MR. MCDADE (O.S.)
> And at the bottom of the page,
> the earlier email: "Laura does
> not want me to attend any of
> the . . .

BACK ON MR. MCDADE

> MR. MCDADE
> . . . sockeye salmon workshops
> that are not run by DFO for
> fear that we will not be . . .

BACK TO EMAIL — Exhibit 628. The words
"control the way the disease issue could be
construed in the press" are highlighted.

>   MR. MCDADE (O.S.)
>   . . . able to control the way
>   the disease issue could be
>   construed in the press."

BACK TO DR. RICHARDS

>   DR. RICHARDS
>   Again, that is also a
>   misrepresentation of what was
>   going on.

>   MR. MCDADE (O.S.)
>   Did you talk to her about this?

>   DR. RICHARDS
>   The issue here is that we were
>   trying to wait to get some
>   more results from her before
>   we put the note up, and then
>   suddenly we had a meeting that
>   was called and we hadn't, and
>   we hadn't had a chance to do
>   the full briefing up the line.
>   So it's not that we were trying
>   to hide something. It was more
>   that we were trying to make
>   sure that we had information go
>   up the line before people were
>   surprised by reports in the
>   media.

BACK ON MR. MCDADE

                    MR. MCDADE
          So once you'd given the
          disease briefing note, then any
          reluctance to let her go to the
          media would have been gone?

                    DR. RICHARDS (O.S.)
          Yes. On my part. But as I say,
          within the department, there's
          a process on this, and this
          process was then complicated by
          calling the formation of this
          Commission.

ON DR. RICHARDS

                    MR. MCDADE (O.S.)
          Dr. Richards, when Dr.
          Miller released her *Science*
          paper . . .

ON MR. MCDADE

                    MR. MCDADE
          . . . in January of this year —

                    DR. RICHARDS (O.S.)
          — Yes.

                    MR. MCDADE
          She was told she should not
          speak to the media. Are you
          aware of that?

                    DR. RICHARDS (O.S.)
        Yes.

PAN TO DR. RICHARDS

                    MR. MCDADE (O.S.)
        Throughout all of 2010, she was
        told she shouldn't speak to the
        media. Isn't that right?

                    DR. RICHARDS
        I cannot, I'm not aware of
        other instances.

ON THE COMMISSIONER

                    MR. MCDADE (O.S.)
        I'd like to take you to Exhibit
        622-A, the speech,

DOCUMENT — Speaking notes of a Member of
Parliament.

                    MR. MCDADE (O.S.)
        . . . the speaking notes for a
        Member of Parliament.

BACK ON MR. MCDADE

                    MR. MCDADE
        Are you aware those speaking
        notes were being produced
        for an emergency session of
        Parliament?

> DR. RICHARDS (O.S.)
> Yes, I gave that evidence
> already today.

> MR. MCDADE
> Right.

BACK TO DOCUMENT — Highlighting "Low returns
of sockeye salmon to the Fraser River."

> MR. MCDADE (O.S.)
> And the title, as I see, is
> "For a Debate on Low Returns of
> Sockeye Salmon to the Fraser
> River."

BACK TO MR. MCDADE

> MR. MCDADE
> That's the, ah, you understood
> that was the subject of the
> speech?

ON DR. RICHARDS

> DR. RICHARDS
> I wasn't party to how that work
> was being divided up.

PAN BACK TO MR. MCDADE

> MR. MCDADE
> Well, Doctor, when I read that
> speech, it seems to be a whole
> bunch about how sea lice is not
> (MORE)

          MR. MCDADE (CONT'D)
the problem but there's not
one word in that speech about
virus. Is that right?

          DR. RICHARDS (O.S.)
As I say, I did not write
these.

ON DR. RICHARDS

          DR. RICHARDS
They were based on information
that we, that we had and some
of that information may have, I
know that what they were doing,
though, was trying to divide up
some of this material between
various different notes. I
think the issue of diseases was
somewhere in these. I don't
recall specifically which one.

BACK TO DOCUMENT — On page 7, the words
"Including the impact of climate change" are
highlighted.

          MR. MCDADE (O.S.)
If we go to page 7 of that
document. Do you see the second
paragraph? A number of factors
could be the cause, including
the impact of climate change.

BACK ON MR. MCDADE

                    MR. MCDADE
          Now, you knew, Dr. Richards, at
          the time that speech was being
          written for Parliament that
          virus was the leading cause,
          likelihood, and why is it that
          it doesn't appear —

OVER TO MR. TAYLOR

                    MR. TAYLOR
          — I object.

                    MR. MCDADE (O.S.)
          Sorry?

                    MR. TAYLOR
          Dr. Richards isn't here to
          answer how someone else wrote
          a briefing note or a speech,
          rather.

BACK ON MR. MCDADE

                    MR. MCDADE
          I think it's a relevant
          question. I have nothing more
          to say about that on that
          matter.

PANNING OVER TO MR. TAYLOR

                    MR. TAYLOR
          Well, in addition, she hasn't
          testified, and I don't think
          Mr. McDade has put forward any
          evidence, that everyone knew
          that a virus was the leading
          cause of something or other.

BACK ON MR. MCDADE

                COMMISSIONER COHEN (O.S.)
          It might be just a question of
          rephrasing your question.

                    MR. MCDADE
          Why, Dr. Richards, didn't you
          ensure in your advice on this
          document that Parliament was
          informed about the virus?

PAN TO DR. RICHARDS

                    DR. RICHARDS
          Well, first of all, as I
          mentioned, these things were
          done through different notes.
          And we were still trying to
          determine whether, in fact,
          there was a virus.

ON MR. MCDADE

                    MR. MCDADE
          Can I go to Tab 22 of my
          documents? Dr. Richards,

DOCUMENT — Close on the heading "Fiscal Year
Jan-Mar 2010."

                    MR. MCDADE (O.S.)
          . . . here's a request from
          Dr. Miller for funding for the
          fiscal year January to March
          2010 . . .

ON THE WORDS — "Identify and sequence the
virus"

                    MR. MCDADE (O.S.)
          . . . to identify the virus in
          the retrovirus family.

BACK ON MR. MCDADE

                    MR. MCDADE
          You'll see that in the middle
          of the page,

BACK TO DOCUMENT — Highlight "87k."

                    MR. MCDADE (O.S.)
          . . . $87,000 in funding is
          requested. Was that provided?

                    DR. RICHARDS (O.S.)
          I don't know.

ON DR. RICHARDS

                    DR. RICHARDS
          Perhaps you should ask Dr.
          Miller about that.

DOCUMENT — Exhibit 638, "Proposed Research on
Suspected Novel Virus from Genomics Study on
Sockeye Salmon."

                    MR. MCDADE (O.S.)
          Can I go to Tab 25? There's
          another funding request. That's
          a funding request to, again,
          establish the prevalence
          and intensity of the viral
          signature.

BACK ON MR. MCDADE

                    MR. MCDADE
          And you'll see the cost is
          set out in bold in the middle
          paragraph.

ON THE PAGE — Push in and highlight "COST
to establish using arrays whether viral-
signature is present in Atlantic salmon
$18,750."

                    MR. MCDADE (O.S.)
          "COST to establish using arrays
          whether viral-signature is
          present in Atlantic salmon
          $18,750."

ON MR. MCDADE

                MR. MCDADE
      Was that funded?

ON DR. RICHARDS

                DR. RICHARDS
      I don't know whether that
      specifically was funded. I
      believe it was but I can't
      confirm that exactly.

ON COMMISSIONER COHEN — He is watching Dr.
Richards.

THEN ON DR. RICHARDS — She is watching her
monitor.

DOCUMENT — Push in and highlight "$23.5
million."

                MR. MCDADE (O.S.)
      The DFO has contributed
      something like $23 million to
      aquaculture research,

ON DR. RICHARDS

                MR. MCDADE (O.S.)
      . . . including making
      genetically modified fish that
      would be protected from virus.
      Why is the department not
      prepared to fund this kind of
      work?

ON MR. TAYLOR

                    MR. TAYLOR
          That's not something for this
          witness.

                    MR. MCDADE (O.S.)
          Well, isn't she the person
          who testified as to how
          funding . . .

BACK TO MR. MCDADE

                    MR. MCDADE
          . . . took place?

Commissioner Cohen is listening intently.

                    DR. RICHARDS (O.S.)
          I mean, I think we would be
          interested in the work.

ON COMMISSIONER COHEN'S REACTION

BACK TO DR. RICHARDS

                    DR. RICHARDS
          I don't, I'm just not sure
          about this particular, whether
          this was in fact done or not.

                    MR. MCDADE (O.S.)
          Thank you. Thank you for the
          extra time, Mr. Wallace.

                                        FADE OUT:

CHAPTER 9

The Question of Sea Lice.

Or,

"Is it what sea lice carry that is the problem?"

Sockeye smolts throughout the narrow channels around the Discovery Islands were often found to have heavy sea lice (*Lepeophtheirus salmonis* or *Caligus clemensi*) loads. These smolts were collected on June 28, 2008, near Cyrus Rock. Gravid salmon louse females (the white egg strings) can produce up to 6,600 eggs per female per year. The salmon louse can complete its life cycle in just 4–9 weeks, depending on the water temperature. This means that sea lice can generate new sea lice several times per year. Although sea lice can cause tremendous physical damage to their hosts, they are now recognized as vectors, or carriers of pathogens, that can transfer pathogens from one host to another. (Photo: Jody Eriksson)

INT. COURTROOM — MORNING

Commissioner Cohen enters and takes his seat.

SUPERSCRIPT: September 6, 2011 . . .

            REGISTRAR GILES
      The hearing is now resumed.

ON MS. GRANT

              MS. GRANT
      Counsel for Canada is up next
      with 30 minutes.

ON MR. TAYLOR

              MR. TAYLOR
      Dr. Saksida, do you have a
      comment about lice as a vector
      of pathogens?

ON DR. SAKSIDA

             DR. SAKSIDA
      Sea lice are more of a
      mechanical vector . . .

LOWER THIRD: Dr. Sonja Saksida, Executive Director, Centre for Aquatic Health Sciences.

             DR. SAKSIDA
    . . . than an actual, true
      vector for transmission of
      disease. It looks like they,
            (MORE)

                    DR. SASKIDA (CONT'D)
          they may, when a motile stage,
          a larger louse is attached to
          a fish, if there's another, and
          they're diseased. If they're
          heavily diseased, this louse
          may actually pick up the
          virus or the bacteria, swim
          to the next host, and there is
          potentially a transmission.

BACK TO MR. TAYLOR

                    MR. TAYLOR
          And turning to you, Dr. Jones,
          and having heard what's been
          said, what's your comment?

ON DR. JONES

**LOWER THIRD: Dr. Simon Jones, Research
Scientist, The Department of Fisheries and
Oceans.**

                    DR. JONES
          The a, the list that was
          referred to in Professor Dill's
          report was actually a list
          of, ah, references to the
          scientific literature where
          researchers had associated a
          particular fish pathogen with
          salmon lice.

ON MR. TAYLOR

> DR. JONES (O.S.)
> For example, with IHN virus
> or with ISA virus, or with
> *Aeromonas salmonicida* . . .

BACK TO DR. JONES

> DR. JONES
> . . . bacterial pathogen in
> salmon, there is evidence
> that these pathogens have been
> associated with the salmon
> louse. But that's a very
> different piece of information
> than, than saying that the
> salmon louse, because of its
> biology and behaviour, is a
> competent vector of those, of
> those pathogens.

OVER TO MR. TAYLOR

> MR. TAYLOR
> Dr. Noakes and Dr. Dill, who
> will be familiar to the
> panellists, gave evidence
> earlier and opined that
> sea lice is unlikely to be
> the cause of the decline in
> productivity of Fraser sockeye,
> although Dr. Dill wasn't as
> certain as Dr. Noakes and
> wouldn't rule it out, but he
> hadn't found any evidence in
> that regard.

ON DR. ORR

>MR. TAYLOR (O.S.)
>Dr. Orr, you agree that sea
>lice is unlikely to be found
>the . . .

ON COMMISSIONER COHEN

>MR. TAYLOR (O.S.)
>. . . cause of decline in
>productivity of Fraser sockeye?

ON DR. ORR

LOWER THIRD: Dr. Craig Orr, Former Executive Director, Watershed Watch.

>DR. ORR
>Well, it's a little difficult
>to take it in isolation. I know
>you want to go there, but I
>think Dr. Dill did suggest that
>he was concerned about it being
>a vector for disease. Is that
>not correct? In terms of how he
>characterized his concern?

>MR. TAYLOR (O.S.)
>He has concerns about
>vectoring, yes.

                    DR. ORR
          Yes. And I would agree with
          Dr. Dill in that case, that is
          something that ah, does need to
          be examined in this Commission.

                                    FADE TO:

ON DR. SAKSIDA

                    MR. MARTLAND (O.S.)
          Counsel for the Aquaculture
          Coalition at 20 minutes.

                    MR. MCDADE (O.S.)
          I took a look at your CV today,

TIGHT ON MR. MCDADE

                    MR. MCDADE
          . . . Dr. Saksida, and it
          seemed as I went down the
          list of projects that you're
          involved in, every single
          one was being funded by the
          industry, isn't that right?

PANNING OVER TO DR. SAKSIDA

                    DR. SAKSIDA
          A large proportion is, yes.

PANNING BACK TO MR. MCDADE

                    MR. MCDADE
          Okay. So you wouldn't describe
          yourself as impartial, to be
          fair, would you?

OVER TO DR. SAKSIDA

                    DR. SAKSIDA
          I think that based on my work
          and my history, I am impartial.
          I think it's very important
          that we have aquaculture. I
          think it's very important that
          it's done properly. And so I am
          critical of the industry when
          it is not done properly, and I
          will praise the industry when
          it is done properly.

OVER TO MR. MCDADE

                    MR. MCDADE
          Now, you also work very
          closely, it seems to me,
          with DFO. For instance, you
          published a number of times
          with Dr. Jones here, right?

ON DR. SAKSIDA

                    DR. SAKSIDA
          I think we have two papers.

                    MR. MCDADE (O.S.)
          Yes.

OVER TO MR. MCDADE

                    MR. MCDADE
          And I think you have two or
          more papers with Dr. Beamish,
          who we've heard from.

                    DR. SAKSIDA (O.S.)
          Yes, we worked on a project
          together through the Pacific
          Salmon Forum.

                    MR. MCDADE
          And Dr. Marty?

                    DR. SAKSIDA (O.S.)
          Dr. Marty and I have worked on
          obviously that other project,
          yes.

                    MR. MCDADE
          And he's a friend of yours?

PAN TO DR. SAKSIDA

                    DR. SAKSIDA
          He's a colleague.

                    MR. MCDADE (O.S.)
          Dr. Kent, you publish with?

                    DR. SAKSIDA
          Michael Kent was my supervisor
          when I was doing my master's.

ON DR. SAKSIDA

                    MR. MCDADE
          So let me, let me ask you this.
          I'm interested in a number of
          studies that you've done around
          sea lice over the last five
          or six or seven years, quite a
          few.

ON DR. SAKSIDA

                    MR. MCDADE (O.S.)
          In effect, that's the primary
          subject matter that you've
          researched and published on in
          the recent past, yes?

                    DR. SAKSIDA
          Um, yes, because that's where
          all the funding was coming
          from. But really,

BACK ON MR. MCDADE

                    DR. SAKSIDA (O.S.)
          . . . Mr. McDade, my job isn't
          to publish. My job is . . .

ON DR. SAKSIDA

DR. SAKSIDA
. . . fish health and fish
welfare, and the publications
just come out because there's
the need to inform.

BACK OVER TO MR. MCDADE

MR. MCDADE
Well, I wouldn't suppose that
you went to school to try and
learn more about sea lice. Why
are you, why are you doing so
many studies on sea lice?

PAN OVER TO DR. SAKSIDA

DR. SAKSIDA
Obviously because the Pacific
Salmon Forum in its wisdom,
when they were trying to
determine the effects of, on
wild fish concentrated on sea
lice.

ON MR. MCDADE

MR. MCDADE
If we could just highlight the
second paragraph. Over the last
few years, you say there, Dr.
Saksida, that . . .

ON DR. SAKSIDA

                    MR. MCDADE (O.S.)
          . . . a large percentage of
          your "time and research efforts
          are spent responding to . . .

BACK TO MR. MCDADE

                    MR. MCDADE
          . . . this debate which has
          become a vocal and often
          reoccurring topic, and the
          negative news stories presented
          by NGOs make great headlines,

ON DR. SAKSIDA

                    MR. MCDADE (O.S.)
          . . . and responses that call
          into question the motives of
          and provide a critical analysis
          do not.

ON MR. MCDADE

                    MR. MCDADE
          Responding to the same
          repeated messages and faulty
          science has become a source
          of frustration for me and many
          others throughout the BC salmon
          farming industry." Is that
          correct, that this has become a
          source of frustration for you?

PANNING OVER TO DR. SAKSIDA

                    DR. SAKSIDA
          I find it very, very difficult
          to handle listening to, um,
          information that I find
          incorrect, and most of the time
          I leave it, sometimes I respond,
          and, yes, I can find it quite
          frustrating.

                              FADE OUT:

CHAPTER 10

The Question of Wild vs. Farmed Fish.

Or,

"Who should study what?"

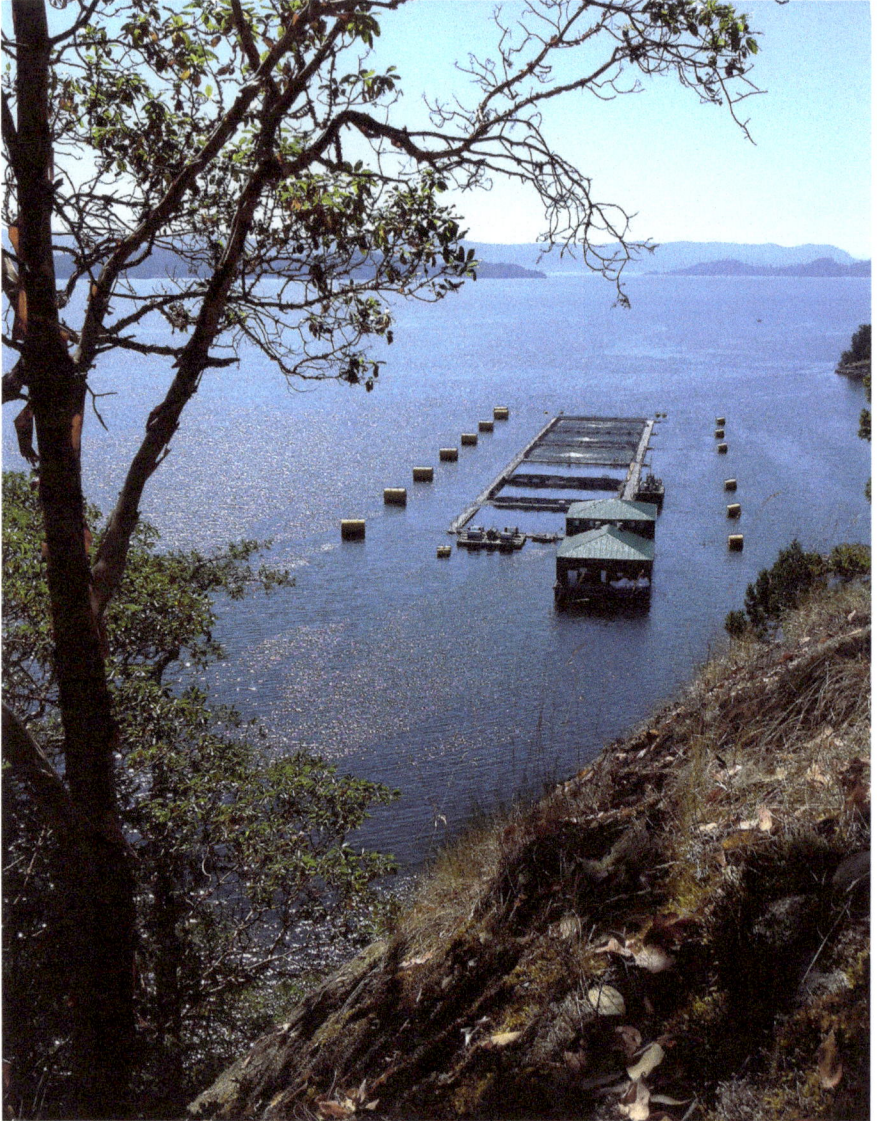

At the peak of the salmon farming frenzy, 130 marine open net pen fish farms operated around the Campbell River, Port Hardy and Tofino areas of Vancouver Island. The Department of Fisheries and Oceans struggles with the dual mandate of supporting the aquaculture industry and protecting wild fish. This has undoubtedly contributed to wild fish populations not being protected from the diseases being amplified in the farms. (Photo: Jody Eriksson)

INT. COURTROOM — DAY

Dr. Kent is on the stand.

SUPERSCRIPT: August 23, 2011 . . .

        MR. MCDADE (O.S.)
Could we have Dr. Kent's CV up
on the screen?

ON MR. MCDADE

        MR. MCDADE
You're primarily an expert in
diseases in fish farms.

        DR. KENT (O.S.)
No. I disagree with that.

PAN TO DR. KENT

        MR. MCDADE (O.S.)
All right. Well, while you were
in BC, that was primarily your —

ON DR. KENT

LOWER THIRD: Dr. Michael Kent, Professor, Oregon State University.

        DR. KENT
— That's correct. —

                    MR. MCDADE (O.S.)
          — expertise. All right. And
          that's the basis upon which
          you've been called to become
          an expert at the Commission, I
          would presume.

                    DR. KENT
          I disagree with that. Actually,
          my conversations with Dave Levy
          were and my CV was twofold,
          why I think I'm appropriate
          for this. One is my past
          experience with DFO working
          with the net pen farms, and my
          present experience working with
          diseases in wild salmonids.

ON MR. MCDADE

                    MR. MCDADE
          Your report didn't cover the
          problems from fish farms,
          regardless of the reason why.

                    DR. KENT (O.S.)
          That's correct.

                    MR. MCDADE
          It didn't, did it?

                    DR. KENT (O.S.)
          It did not.

BACK TO DR. KENT

                    DR. KENT
          I did not talk about what the
          role of fish farms would be
          in transmitting it to sockeye
          salmon.

PAN OVER TO MR. MCDADE

                    MR. MCDADE
          Okay. And this reviewer's
          comment, you'll agree, was that
          you should have.

                    DR. KENT (O.S.)
          That's his comment, yes.

                    MR. MCDADE
          And your answer is in the bold,
          there, at the top of page 56.

DOCUMENT — page 56. The words "Fish farms
and sea lice are dealt with in more depth in
another report (Report 5)" are highlighted.

                    MR. MCDADE (O.S.)
          "Fish farms and sea lice are
          dealt with in more depth in
          another report."

                    DR. KENT (O.S.)
          That's correct.

BACK TO MR MCDADE

                    MR. MCDADE
          And which report is that?

                    DR. KENT (O.S.)
          That's with the various fish
          farm, the Report 5, and at
          that . . .

PAN OVER TO DR. KENT

                    DR. KENT
          . . . now I realize there are
          several reports that are coming
          out on, on fish farms, so
          that's where it was being dealt
          with.

                    MR. MCDADE (O.S.)
          But, but you were the disease
          expert contracted to deal with
          these questions, and they're
          not disease experts, are they?

                    DR. KENT
          I don't know their expertise.

OVER HIS SHOULDER — Commissioner Cohen is
listening intently.

                    MR. MCDADE (O.S.)
          Well, before this review ever
          came in, you'd already decided
          consciously to . . .

ON MR. MCDADE

                    MR. MCDADE
     . . . ignore fish farms, hadn't
     you?

                    DR. KENT (O.S.)
     No.

                    MR. MCDADE
     But you didn't do it.

PAN BACK TO DR. KENT

                    DR. KENT
     That's right, because I did not
     find great evidence of diseases
     being transmitted from fish
     farms in being in the role of
     sockeye salmon.

ON MR. MCDADE

                    MR. MCDADE
     Did you look at the Fish Health
     Database?

WHIP TO DR. KENT

                    DR. KENT
     Which exhibit is that one?

                    MR. MCDADE (O.S.)
     Um, Mr. Lunn, could we have up
     the list of documents that I
     referred to as the Fish Health
     Database?

INSERT — Exhibit 1505. Scrolling down the spreadsheet.

> MR. MCDADE (O.S.)
> Dr. Kent, that's the actual spreadsheets and . . .

BACK ON MR. MCDADE

> MR. MCDADE
> . . . reports of the fish health auditing and the reports that the fish farms make to the Province around fish health. Did you look inside those documents?

BACK TO DR. KENT

> DR. KENT
> Yeah, I've looked at them. They came to me quite late. I actually reviewed them this morning. I scanned through them. I, I scanned through them, they're pretty extensive, but I didn't go through them in all sorts of detail.

PANNING OVER TO MR. MCDADE

> MR. MCDADE
> So did you see, did you have them when you did your report?

> DR. KENT (O.S.)
> No, I didn't.

> MR. MCDADE
> Well, wouldn't they be relevant
> to your report if there's
> diseases that are all over
> those spreadsheets?

PAN TO DR. KENT

> DR. KENT
> They'd be useful. It's not
> peer-reviewed literature, but
> they would be useful.

> MR. MCDADE (O.S.)
> What's, what's the distinction
> from peer-reviewed literature?

> DR. KENT
> It's been validated by
> professionals. It would be
> of use, but I, I, given the
> limitations that I had with my
> time, the most useful data were
> peer-reviewed papers for the
> study.

PAN BACK TO MR. MCDADE

                    MR. MCDADE
          And so if DFO hasn't studied
          a matter, if there's no peer-
          reviewed paper on it, for you,
          it didn't exist?

                    DR. KENT (O.S.)
          No, I said it has less
          significance to me.

ON DR. KENT

                    MR. MCDADE (O.S.)
          So you didn't look at how many
          times the disease marine anemia
          has been diagnosed in BC fish
          farms over the last 10 years.

                    DR. KENT
          No, I didn't, and I don't see,
          and I don't see a diagnosis of
          marine anemia on here.

BACK ON MR. MCDADE

                    MR. MCDADE
          No. But that's my point. You
          don't know how many times
          marine anemia has been
          diagnosed, do you?

                    DR. KENT (O.S.)
          No, I don't.

> MR. MCDADE
> You don't know how many times
> IHN has been diagnosed, do you?

POP-UP — IHN: Infectious Hematopoietic Necrosis

> DR. KENT (O.S.)
> That, that information is from
> other . . .

PAN BACK TO DR. KENT

> DR. KENT
> . . . reports, et cetera,
> where, where other, other data
> report that there have been no
> outbreaks of IHN. That, that's
> from other grey literature data
> that were given to me.

ON MR. MCDADE

> DR. KENT (O.S.)
> So I'm not relying on, on . . .

BACK TO DR. KENT

> DR. KENT
> . . . the absence of IHN
> outbreaks in, in, in BC farms,
> based on this database. I was
> basing that on summaries from
> other documents that I had
> (MORE)

>   DR. KENT (CONT'D)
> available for me when I was
> preparing this, when I was
> preparing this, this overview.

BACK TO MR. MCDADE

>   MR. MCDADE
> Summaries of other documents,
> that is, something that the
> people who prepared these
> documents have summarized for
> you?

ON DR. KENT

>   DR. KENT
> That's my understanding.

>   MR. MCDADE (O.S.)
> So you had to rely entirely
> on what you were told about
> whether these diseases
> appeared.

>   DR. KENT
> And the peer-reviewed
> literature.

ON THE CROWD

>   MR. MCDADE (O.S.)
> Fish farms can get a disease
> from wild fish . . .

BACK ON MR. MCDADE

          MR. MCDADE
. . . and then incubate it or
amplify it, can't they?

          DR. KENT (O.S.)
That's correct.

          MR. MCDADE
Fish farms can take a disease
that's present in an avirulent
form in a, in wild fish,
and . . .

ON DR. KENT

          MR. MCDADE (O.S.)
. . . have it mutate to a
virulent form. That's been seen
as well, hasn't it?

BACK ON MR. MCDADE

          DR. KENT (O.S.)
Virulence in a fish farm from a
wild —

PAN TO DR. KENT

          DR. KENT
No, I don't know of a specific
example of that.

ON DR. STEPHEN

                    MR. MCDADE (O.S.)
            Dr. Stephen, would you agree
            with that comment?

ON COMMISSIONER COHEN

                    DR. STEPHEN (O.S.)
            Oh, I'm not aware of a case of
            that either.

                    MR. MCDADE (O.S.)
            What about ISA in Norway? No?

ON DR. KENT

                    DR. KENT
            Actually, Dr. MacWilliams has
            done, who has done a lot of
            work on ISA, maybe she could
            respond to that.

PAN TO DR. MACWILLIAMS

LOWER THIRD: Dr. Christine MacWilliams, Fish Health Veterinarian, Department of Fisheries and Oceans.

                    DR. MACWILLIAMS
            There is a recent publication
            that proposes that that has
            happened, that the avirulent
            form may have in a farm mutated
            to a virulent form of ISA.

POP-UP — ISA: Infectious Salmon Anemia

BACK ON MR. MCDADE

> MR. MCDADE
> ISA, pancreas disease,
> HSMI . . .

POP-UP — HSMI: Heart and Skeletal Muscle Inflammation

> MR. MCDADE
> . . . were never heard of in wild fish in Norway until they had fish farms, right?

PAN TO DR. KENT

> DR. KENT
> That's correct. That's just what I was discussing.

> MR. MCDADE (O.S.)
> And if a disease like HSMI shows up in Canada,

ON THE REPORT — Push in on the heading HSMI and highlight "Heart and skeletal muscle inflammation."

> MR. MCDADE (O.S.)
> . . . your conclusion would likely be that it came from wild fish?

> DR. KENT (O.S.)
> Yes.

NEXT PAGE — Highlight ISA and push in to see "Infectious Salmon Anemia."

> MR. MCDADE (O.S.)
> Over the page, please. The same with the section on ISA. If ISA shows up in Canada, are you going to conclude that it came from wild fish?

PAN BACK TO DR. KENT

> DR. KENT
> Yeah, I, I, from based on what I know, and Dr. MacWilliams might be able to expand on that, 'cuz she's an expert on ISA, my understanding of ISA when they do the genetic typing, when it moved from Scotland, when they observed it first in Norway, and then they saw it in Scotland, the, the first assumption was that, oh, it must have come with fish farming activities between Norway and Scotland. Genetic typing of the virus showed that they were quite distinctive and it was confined to the marine environment. And actually they,

ON MR. MCDADE

> DR. KENT (O.S.)
> . . . they occurred
> independently in Norway and
> Scotland.

> DR. MACWILLIAMS (O.S.)
> I would say that I would
> disagree.

OVER TO DR. MACWILLIAMS

> DR. MACWILLIAMS
> If ISA were detected here, I
> would presume it came from a
> break in biosecurity, either
> at a farm level or through
> international transport. I
> would not presume it's coming
> from wild fish in BC, because
> there have been tests, and
> people have looked for ISA with
> very sensitive micro tests and
> it has not been found. So I
> would presume that that was an
> iatrogenic introduction, that a
> break in biosecurity somewhere
> along the line.

> FADE TO:

INT. COURTROOM — DAY

Mr. McDade is at his podium.

                    MR. MCDADE
          Dr. Stephen, let me turn to
          you and a couple papers you
          wrote . . .

ON DR. STEPHEN

                    MR. MCDADE (O.S.)
          . . . in the '90s on the marine
          anemia. So this is referring,
          more . . .

BACK ON MR. MCDADE

                    MR. MCDADE
          . . . or less, to the same
          phenomena: "Prior to the
          project, marine anemia had not
          been diagnosed on 15 of the 23
          farms we visited."

                    DR. STEPHEN (O.S.)
          I'd have to recheck those
          numbers, but yes, there it is,
          that's correct, yep.

                    MR. MCDADE
          "And we later found cases of
          marine anemia in all but 1 of
          the 23 farms."

PAN TO DR. STEPHEN

LOWER THIRD: Dr. Craig Stephen, Professor, Faculty of Veterinary Medicine, University of Calgary.

                    DR. STEPHEN
          Yes. Anywhere we could find a
          farm that had chronic, other
          chronic inflammatory diseases,
          we could find one or more cases
          of this disease.

                    MR. MCDADE (O.S.)
          And it says, "On average," a
          couple of lines down. "On
          average, 2 visits per site were
          required before the disease was
          diagnosed."

                    DR. STEPHEN
          Correct.

ON MR. MCDADE

                    MR. MCDADE
          Right. So as I understood
          it, there's a disease that's
          killing salmon in 1988. You
          do a study in 1990 that calls
          it plasmacytoid leukemia. Your
          survey in 1991 finds that
          sockeye are highly susceptible
          to it. And you look at chinook
          in 1994. You don't look at
          sockeye until 1998. Can you
          explain to me why you didn't
          do anything about this disease
          in relation to the wild sockeye
          for 10 years?

PAN TO DR. KENT

                    DR. KENT
          It wasn't 10 years, but the
          reason why is we, we recognized
          this as a disease of chinook
          salmon, so it was a disease
          of chinook salmon. So if we're
          going to look at . . .

ON MR. MCDADE — He looks at his notes.

                    DR. KENT (O.S.)
          . . . fish in the wild, the
          first fish we would look at
          would be chinook salmon.

ON THE GALLERY — It's half full.

ON MR. MCDADE

                    MR. MCDADE
          You didn't look, as we
          understand it, we've heard
          other evidence, that in the
          early '90s sockeye salmon were
          starting to show abnormal early
          entry behaviour into the Fraser
          that was affecting pre-spawn
          mortality. You've heard about
          that?

                    DR. KENT (O.S.)
          In the early '90s?

                    MR. MCDADE
          Yes.

                    DR. KENT (O.S.)
          Yeah, I've heard about that, of
          course.

                    MR. MCDADE
          Did you, at that time, connect
          those two events in any way?

                    DR. KENT (O.S.)
          No.

                    MR. MCDADE
          Were you, in the marine, the,
          the aquatic health branch, ever
          consulted about the expansion
          of the fish farm industry . . .

ON DR. KENT

                    MR. MCDADE (O.S.)
          . . . in relation to disease?

                    DR. KENT
          Yes.

BACK TO MR. MCDADE

                    MR. MCDADE
          Did you discuss this early
          entry problem and the
          possibility that salmon
          leukemia virus was behind it?

                    DR. KENT (O.S.)
          No.

                    MR. MCDADE
          Never occurred to you?

                         DR. KENT (O.S.)
          No.

                    MR. MCDADE
          So today, if DFO finds a new
          virus that they haven't seen,
          is there anything different
          that's happening at DFO that
          would take less than 10 years
          to discover the impacts on the
          wild salmon?

OVER TO MR. TAYLOR

                    MR. TAYLOR
          I object. This witness isn't
          there. He's already testified
          he left in 1999.

ON THE PANEL

                    MR. MCDADE (O.S.)
          Fair enough. Let me ask that
          question of Dr. Johnson. Is
          there anything that would take
          less than 10 years to determine
          this kind of impact?

BACK ON MR. MCDADE

                         DR. JOHNSON (O.S.)
          Yes.

ON DR. JOHNSON

LOWER THIRD: Dr. Stewart Johnson, Aquatic Animal Health, Department of Fisheries and Oceans.

> DR. JOHNSON
> For example, we are conducting challenge trials with a virus which was recently identified in sockeye salmon.

> MR. MCDADE (O.S.)
> Yes, but have you done, did you do anything in terms of getting the fish farms out of the path of the migratory salmon, or did you just do more studies?

> DR. JOHNSON
> I, personally, don't do anything about getting the fish farms out of the migratory path, because that makes the assumption that the fish farms are a significant source of pathogens.

> MR. MCDADE (O.S.)
> So DFO wouldn't react until there was scientific proof that connected the fish farms and the pathogens and the harm to the wild sockeye?

ON MR. MCDADE

>   DR. JOHNSON (O.S.)
> I can't answer how senior
> management would react.

ON DR. JOHNSON

>   DR. JOHNSON
> It's a question you need to ask
> of the senior managers of DFO.

>   MR. MCDADE (O.S.)
> All right. Well, perhaps we
> will.

                                        FADE OUT:

CHAPTER 11

The Question of Exotic Viruses.

Or,

"Is it not okay to talk

about some things?"

This female pink salmon (*Oncorhynchus gorbuscha*) was photographed on September 10, 2015, at High Falls Creek, a tributary of the Squamish River. She was surrounded by five males that were keen to spawn with her because there was a noticeable imbalance between the numbers of males and females spawning at that time. She tried several times to dig a redd, but suddenly turned belly-up and died. The bleeding fin and red lesions are common symptoms of viral hemorrhagic septicemia (VHS). Many other pink salmon carcasses at this location had similar symptoms. The Canadian Food Inspection Agency (CFIA) prefers that Canadian pathologists do not test for suspected international reportable diseases that could have a negative impact on trade. This appears to be a common practice in many countries. (Photo: Scott Renyard)

INT. COURTROOM — DAY

Commissioner Cohen enters.

SUPERSCRIPT: August 31, 2011 . . .

Commissioner Cohen bows to the court.

ON REGISTRAR GILES

>   REGISTRAR GILES
>   The hearing is now resumed.

>   DISSOLVE TO:

**LOWER THIRD: Jonah Spiegelman, Associate
Counsel, Government of Canada.**

>   MR. SPIEGELMAN
>   Dr. Marty, are you aware
>   that Commission counsel in
>   these proceedings permitted
>   Alexandra Morton to make a
>   confidential report to the CFIA
>   regarding documents she found
>   in the Commission's disclosure
>   database?

PANNING TO DR. MARTY

>   DR. MARTY
>   My interpretation of the actual
>   order is a little unclear, but
>   I know that the report did
>   occur.

LOWER THIRD: Dr. Gary Marty, Fish Pathologist, Animal Health Centre, BC Ministry of Agriculture.

PAN BACK AND FORTH

> MR. SPIEGELMAN
> And how do you know about that?

> DR. MARTY
> I was informed about it through, I think, reading one of Alexandra Morton's blogs, or possibly from CFIA. I don't remember exactly.

BACK TO MR. SPIEGELMAN

> MR. SPIEGELMAN
> Okay. And the documents at issue were, were reports that you authored. Is that correct?

> DR. MARTY (O.S.)
> Yes.

INSERT — Exhibit 1666 and turn to page 3 and the line "All cases were evaluated as NO RISK for ISA."

> MR. SPIEGELMAN (O.S.)
> On page 3 of this document, it states that "All cases were evaluated as NO RISK for ISA."
> (MORE)

                    MR. SPIEGELMAN (CONT'D)
          Dr. Marty, is that evaluation
          consistent with the conclusions
          you . . .

ON COMMISSIONER COHEN

                    MR. SPIEGELMAN (O.S.)
          . . . reached as to these
          particular cases in the first
          place?

                    DR. MARTY (O.S.)
          Yes, I have seen that before.

BACK TO MR. SPIEGELMAN

                    MR. SPIEGELMAN
          And that's consistent with your
          original, with your original
          interpretation of the results
          in these cases?

PAN TO DR. MARTY

                    DR. MARTY
          Yes.

                    MR. SPIEGELMAN (O.S.)
          Then if I can ask, why, in
          those cases, did you make
          reference to ISA?

                    DR. MARTY
          Part of my role as a
          pathologist is to provide
          information to my clients.

EXT. HARRISON RIVER — DAY

A dead sockeye floats in a back eddy.

                    DR. MARTY (O.S.)
          In the case of these reports,
          often I would say that viral
          hemorrhagic septicemia . . .

TWO SHOTS — of sockeye with lesions.

                    DR. MARTY (O.S.)
          . . . virus is the most common
          identified cause of these
          lesions of concern.

INT. COURTROOM — MOMENTS LATER

Dr. Marty looks at Commissioner Cohen.

                    DR. MARTY
          But I'm also aware of the
          interest and the potential for
          ISA to come into BC. And so
          in all of these cases I have
          a standard comment that I use
          with this lesion that says
          something like "Sinusoidal
          congestion," which is the
                    (MORE)

                    DR. MARTY (CONT'D)
          lesion of concern, "is a
          classic lesion associated with
          ISAV." That's just a statement
          of fact that provides my
          clients with information. And
          I also include a clause after
          that,

BACK TO DOCUMENT — Exhibit 1666. The line
"ISA has never been reported in British
Columbia" is highlighted.

                    DR. MARTY (O.S.)
          "but ISAV has not been" —
          "never been identified in
          British Columbia."

BACK TO DR. MARTY

                    DR. MARTY
          So in several of these cases,
          it's not routine, when you have
          that level of confidence. It's
          not routine to always test for
          it when it's not known to occur,
          especially when you always have
          this active audit program going
          on. In fact, CFIA actually
          discourages us to test for
          international foreign animal
          diseases. They prefer that they
          be, they be called.

                                        FADE OUT:

CHAPTER 12

The Question of Disease Data.

Or,

"What do they mean by 'Other'?"

Legend: Other · Predators · Environment · Fresh Silver

Chart: y = -52192x + 1E+09 ; R² = 0.6848029 ; Linear (Series1)
Y-axis: Number of Dead Fish ; X-axis: 2002 2003 2004 2005 2006 2007 2008 2009 2010 2011
Secondary chart axis: Millions (3–9) ; Year: 2004 2005 2006
Note: Few farms included in industry database

Sum of Mortality

Row Labels	Background Mortality	Culls / Quality Control	Environmental	Fresh	Fresh "Silvers"	Handling / Transport	Matures	Old	Poor Performers	Predators	Systems Related
2000					30354	2602	4224	16758	2122	6370	
2001	6155				31895	6958	1877	11551	7136	5669	
2002	14869		2	398743	647246	26416	23780	100301	73001	26165	
2003				2584714	3559074	469049	290169	1250995	518204	115794	
2004				733744	811069	386056	45031	621220	712606	131281	
2005	34212	156832		623031	1118583	243298	116160	531645	646406	162479	4996
2006	348539	456677		664201	849297	135676	33928	668948	401243	169642	650
2007				510087	589989	232111	33565	557304	302667	222969	
2008				1436440	624015	427841	88121	620406	775933	216894	12883
2009				517388	696832	250476	39610	535535	474056	209141	56
2010				169469	376181	283510	13442	272535	271815	136617	

Total Mortality by Type

Year	Fresh Silvers	Environmental	Predato Other	Total	% Mortality	Fresh "Silvers"	% Fresh Silvers	N. Mort Fresh Silver	% Mort Fresh Silver
2000	30354		0 4570	65530	11%	30354	50%	5.4%	
2001	31895		0 5669	64826	11%	31895	49%	5.4%	
2002	710054		39826 26155	394249	9%	647246	46%	4.0%	
2003	3617782	2584714		2608235	81+06	1716913	43%	12.7%	
2004	811069	733744		1716913	81+06		24%	2.9%	
2005	1131805	623031	86079	1357220	81+06		36%	3.8%	
2006	849297	664201		1434652	81+06		18%	2.4%	
2007	589989	510087		1968800	81+06		18%	2.8%	
2008	624015	1436440		1862301	41+06		18%	2.0%	
2009	696832	517388		1289985	31+06		26%	2.3%	
2010	376181	169469		841202	21+06		25%	1.3%	
>=2003 avg	1,570,043	903,673		635,096	0.21		22%	2.0%	

Roughly how many farms reported: Fresh "Silvers"

Count of Mortality

Row Labels	Jan	Feb	Mar	Apr	May	Jun	July	Jul	Aug	Sep	Oct	Nov	Dec
2001	12	12	13	13	13	17	15	16	18	10	15	12	13
2002	11	11	12		13	14	10		14				
2003	82	80		62	87	69		85	84	79	76	77	83
2004	75	78		77	80	84		86	86	73	73	81	76
2005	53	53		55	104	101		102	95	90	94	91	91
2006	88	97		94	121	101		95	97	95	97	101	99
2007	57	94		101	107	103		107	102	101	102	101	92
2008	94	96		96	100	101		96	97	88	97	96	96
2009	92	85		98	91	90		88	85	82	85	84	82
2010	67	64		68	76	74		68	70	67			

Sheet tabs: Correlations | ProductionTrend | Population_Data | Population_Summary | Mortality_Summary | Mortality | Mortality Summary | FHI_BCSFA | FHI_BCSFA_Summary | BCMAL_Audit_DX

The interpretation of data can sometimes vary considerably from one researcher to another. This table (Exhibit !544) identifies dead or dying fish in various categories, with one of the largest subsets of dead fish called "other." This document was a key part of the testimony during the disease hearings. (Source: Commission of Inquiry into the Decline of Sockeye Salmon in the Fraser River (Canada), The Uncertain Future of Fraser River Sockeye, Bruce I. Cohen, Commissioner.)

INT. COURTROOM — DAY

Commissioner Cohen enters and bows.

SUPERSCRIPT: August 29, 2011 . . .

Mr. McDade steps into frame.

                    MR. MCDADE
          Let us turn, Dr. Korman, to the
          issue of the fresh silvers.

ON DR. KORMAN

                    MR. MCDADE (O.S.)
          You used fresh silvers as if
          they were the maximum amount of
          disease . . .

BACK ON MR. MCDADE

                    MR. MCDADE
          . . . mortalities that were
          occurring in these farms, did
          you not?

PAN TO DR. KORMAN

                    DR. KORMAN
          Yes.

BACK ON MR. MCDADE

                    MR. MCDADE
          And, um, I think, Dr. Noakes,
          I heard you in your testimony
          last week saying that in your
          view those were the only fish
          that could be, have died of
          disease.

ON DR. NOAKES

**LOWER THIRD: Dr. Don Noakes, Professor,
Mathematics and Statistics, Thompson Rivers
University.**

                    DR. NOAKES
          As I think we said just a few
          minutes ago, you have a much
          higher likelihood of detecting
          a disease in the fish that have
          just died rather than fish that
          are in the pen. And of course,
          I think you probably heard
          Dr. Kent explain that testing
          a fish and finding a pathogen
          doesn't necessarily mean that
          the fish is diseased in terms
          of pathology.

BACK ON MR. MCDADE

                    MR. MCDADE
          Well, I just, I just want to
          get to the assumptions that
          were behind your report, Dr.
          Korman.

ON DR. KORMAN — for a beat.

>   MR. MCDADE (O.S.)
>   So, the two percent number,

INSERT — Exhibit 1544. "2%" is highlighted.

>   MR. MCDADE (O.S.)
>   . . . which is at line 32,
>   it looks like, is the fresh
>   silvers . . .

BACK ON MR. MCDADE

>   MR. MCDADE
>   . . . divided by the total
>   mortalit — the total
>   population, right?

ON DR. KORMAN

>   DR. KORMAN
>   Yeah.

>   MR. MCDADE (O.S.)
>   The biggest total of these four
>   lines, "Fresh Silvers,"

BACK TO TABLE — Exhibit 1544. The other
column on the table is highlighted.

>   MR. MCDADE (O.S.)
>   "Environmental," "Predators"
>   and "Other" is "Other."

ON MR. MCDADE

                    MR. MCDADE
          There's some —

                 DR. KORMAN (O.S.)
          — Right, which, yes, correct.

                    MR. MCDADE
          And "Other" included what, Dr.
          Korman?

BACK TO DR. KORMAN

                    DR. KORMAN
          I'm not exactly sure of the
          breakdown. I never looked
          at it. A big factor would be
          unknown.

ON MR. MCDADE

                    MR. MCDADE
          The largest portion of "Other,"
          as I could see it, in my
          analysis was under the "Old"
          category.

                 DR. KORMAN (O.S.)
          Yes, looks that way.

                    MR. MCDADE
          Now, old, the "Old" category,
          what was your understanding of
          it?

PAN OVER TO DR. KORMAN

>   MR. MCDADE (O.S.)
>   'Cuz fish farm fish don't die
>   of old age.

The people in the gallery laugh.

>   DR. KORMAN
>   Yeah, I'm not, I'm not super
>   familiar with the exact detail
>   of how they classify it as
>   "Old" versus something else.

LOWER THIRD: Dr. Josh Korman, Fish Ecologist, Ecometric Research Inc.

>   MR. MCDADE (O.S.)
>   Let me suggest this to you, Dr.
>   Korman.

PAN TO MR. MCDADE

>   MR. MCDADE
>   The term "Fresh Silver" is used
>   for a fish that's recently died
>   and is floating belly-up.

ON DR. KORMAN

>   DR. KORMAN
>   Correct.

BACK TO MR. MCDADE

                    MR. MCDADE
          The only difference between old
          and fresh silver is when you
          collect them, isn't that right?

ON DR. KORMAN

                    DR. KORMAN
          That makes sense to me, but
          I would like to see the
          documentation on that. I can't
          confirm that.

BACK TO MR. MCDADE

                    MR. MCDADE
          Yes. And so the old would die
          of exactly the same proportion
          of disease as the fresh
          silvers. There's no distinction
          in terms of their cause of
          death, is there?

PAN TO DR. KORMAN

                    DR. KORMAN
          No, so what's the percentage
          difference here, if you add in,
          you've likely done this, if you
          add in the old, does it change
          the numbers a lot?

                    MR. MCDADE (O.S.)
          It does.

                    DR. KORMAN
          The percentage wise?

                         MR. MCDADE (O.S.)
          Yes.

ON THE TABLE — The "Old" category lights up
and the total 5,249,809 appears in red.

                         MR. MCDADE (O.S.)
          There's over five million fish
          in the "Old" category.

                    DR. KORMAN (O.S.)
          Per year.

                         MR. MCDADE (O.S.)
          Your total of that column is
          5.2 million, I suggest to you.

                    DR. KORMAN (O.S.)
          Right.

ON DR. KORMAN

                         MR. MCDADE (O.S.)
          The total number of fresh
          silvers are nine million. The
          total number of old, they're
          five point something million.

                    DR. KORMAN
          So it would change that two
          percent number to —

> MR. MCDADE (O.S.)
> — Three and something.

> DR. KORMAN
> Yeah.

> MR. MCDADE (O.S.)
> Okay.

> DR. KORMAN
> Okay.

> MR. MCDADE (O.S.)
> What about poor performers?

PANNING OVER TO MR. MCDADE

> MR. MCDADE
> Do you know what that heading
> "Poor Performers" means?

> DR. KORMAN (O.S.)
> Yeah, I imagine fish that
> weren't thriving, that weren't
> growing well.

ON DR. KORMAN

> MR. MCDADE (O.S.)
> Well, it's reasonably likely
> to . . .

BACK TO MR. MCDADE

                    MR. MCDADE
          . . . expect the sick would be
          poor performers, isn't it?

PANNING OVER TO DR. KORMAN

                    DR. KORMAN
          No, I could also assume that
          some fish, you know, don't
          jump for pellets, and therefore
          don't thrive in high-density
          conditions and it could have
          nothing to do with disease,
          actually. But that would be a,
          it may be a combination of the
          two.

PAN BACK ON MR. MCDADE

                    MR. MCDADE
          This assumption that you and
          Dr. Noakes made, the fresh
          silvers are the only fish that
          are dying from disease, is a
          mistaken assumption, isn't it?

PAN BACK TO DR. KORMAN

                    DR. KORMAN
          It should be looked at and
          questioned, and that's
          legitimate that you're doing
          that. I don't think it's fair
          to say that all old fish or all
                    (MORE)

                    DR. KORMAN (CONT'D)
          poor performers died of disease
          at all. But I do, I do agree
          with your argument that the
          percentage could be larger than
          what's in the report.

                    MR. MCDADE (O.S.)
          Fair enough.

BACK TO MR. MCDADE

                    DR. KORMAN (O.S.)
          So they're all estimates.

ON THE PANEL'S REACTIONS

                    MR. MCDADE (O.S.)
          So if we add in those poor
          performers,

BACK TO MR. MCDADE

                    MR. MCDADE
          . . . and add them to the
          fresh silvers, the number that
          are dead or possibly dead
          of disease doubles, from two
          percent to four percent.

PAN BACK TO DR. KORMAN

                    DR. KORMAN
          Just glancing at this
          spreadsheet I could see that
          being possible. And then you'd
          have a set of assumptions in
          there with the caveat that
          all old fish and all poor
          performers are assumed to have
          died from disease, as are fresh
          silvers.

COMMISSIONER COHEN — looks at the document.

                    MR. MCDADE (O.S.)
          Let's come to "Environmental,"
          because that's the next major
          category.

BACK ON MR. MCDADE

                    MR. MCDADE
          These, and I should point out,
          these are all self-reported
          headings from the fish farms,
          right?

                    DR. KORMAN (O.S.)
          Yes.

BRIEFLY ON DR. KORMAN

                    MR. MCDADE (O.S.)
          And that may vary from farm to
          farm . . .

ON MR. MCDADE

                    MR. MCDADE
          . . . in terms of what category
          you put something in.

BACK TO DR. KORMAN

                    DR. KORMAN
          There could be some biases
          potentially going in there,
          although they do have to answer
          to the audit to some extent
          in terms of, in terms of their
          farm-level disease.

PANNING BACK TO MR. MCDADE

                    MR. MCDADE
          I'm not suggesting anybody's
          misleading.

                    DR. KORMAN (O.S.)
          Okay.

                    MR. MCDADE
          What I'm suggesting is, if you
          have a bunch of people putting
          something in a number of
          categories, there's a lot
          of subjectivity as to which
          category it gets put into.

                    DR. KORMAN (O.S.)
          That, yes, that seems that way.

ON DR. KORMAN

          MR. MCDADE (O.S.)
This absolute number two
percent.

BACK ON MR. MCDADE

          MR. MCDADE
Dr. Noakes, in your report
you seem to suggest that two
percent was low. And that's a
subjective opinion, isn't it?

PANNING OVER TO DR NOAKES

          DR. NOAKES
Yes, two percent, ah, generally
that would, that would seem
pretty low.

          MR. MCDADE (O.S.)
Well, two percent a year of
disease . . .

BACK ON MR. MCDADE

          MR. MCDADE
. . . or death, and we're not
talking about disease, we're
talking about death from
disease, potentially, if in a
population that is regularly
fed, that's protected from
predators, that seems quite
high to me.

ON DR. NOAKES

                    DR. NOAKES
          Two percent compared to, say, a
          three percent mortality of wild
          fish per day.

BACK TO MR. MCDADE

                    MR. MCDADE
          Aren't we comparing apples and
          oranges there? Because the
          wild fish die from predation
          and looking for food, not from
          disease.

BACK TO DR. NOAKES

                    DR. NOAKES
          Again, but two percent to me
          seems to be low in terms of an
          annual mortality rate.

PANNING OVER TO MR. MCDADE

                    MR. MCDADE
          Well, let me suggest this to
          you. I looked up the Spanish
          flu on, on Wikipedia last
          night. Spanish flu killed 80
          million people and that was
          two point something percent
          of the population. And that's
          considered one of the greatest
                    (MORE)

                    MR. MCDADE (CONT'D)
          epidemics in our, in our history.
          That's a very extraordinary
          amount of death, isn't it, for a
          disease?

PANNING OVER TO DR. NOAKES

                    DR. NOAKES
          That's true in terms of human
          populations. Um, I guess, I
          can't recall, did the Spanish
          flu run over, was it a one-year
          event or?

AND BACK TO MR. MCDADE

                    MR. MCDADE
          Okay. Well, if it's, if it's
          two years, let's call it four
          percent over two years. Right?
          That's what, that's what you
          said was low.

                    DR. NOAKES (O.S.)
          I said two percent was low.

ON DR. NOAKES

                    MR. MCDADE (O.S.)
          Right. And if it's, now that
          we've added up the numbers in
          a different way, and it may be
          four or six percent, you'd have
          to double that, wouldn't you?

                    DR. NOAKES
I mean, that's an upper limit.
And as Josh was, is pointing
out, I mean, the whole thing
has to be ground-truthed in
terms of your fish, your fish
disease checks.

PANNING BACK OVER TO MR. MCDADE

                    MR. MCDADE
You're putting a lot in there
on that, that the fresh silvers
had no disease, in your view,
but they're all dead fish,
aren't they?

                    DR. NOAKES (O.S.)
I'm putting a lot of confidence
in the people who are actually
doing that monitoring in
terms of being able to make
the proper diagnosis, because
they're the experts.

ON DR. KORMAN

                    MR. MCDADE (O.S.)
Dr. Noakes, I take it you'd
agree, that when for . . .

PANNING OVER TO DR. NOAKES

                    MR. MCDADE (O.S.)
          . . . most fish diseases
          there's fewer fish that die
          than fish that get infected and
          fish that get sick.

                    DR. NOAKES
          Certainly there are fish that
          have disease that die of other
          causes.

PANNING OVER TO MR. MCDADE

                    MR. MCDADE
          Well, in the Spanish flu
          example that I gave you, two
          percent died, 28 percent were
          infected. That's a fairly
          standard split in disease,
          wouldn't it be?

BACK TO DR. NOAKES

                    DR. NOAKES
          I don't know what the standard
          is, but there are certainly
          fish that contract a disease
          and survive it, and there is a
          fraction that dies, as well.

BACK TO MR. MCDADE

                    MR. MCDADE
        But if, when you said, two
        percent was low, if in fact 30
        percent of the fish were sick
        or had the pathogen and were
        shedding pathogens, that would
        be high, wouldn't it?

REACTION FROM DR. NOAKES

PAN OVER TO DR. KORMAN

                    DR. KORMAN
        So just to comment, I mean, I
        agree with your, one of your
        premises of your argument is
        that the level of disease as
        far as risk to wild fish could
        be higher than what these
        percentages are because there
        could be a bunch of fish, for
        one thing, that didn't die
        that have the, that have the
        pathogen, right, and therefore
        the risk to wild fish is
        greater than what these
        numbers, ah, suggest.

PAN BACK TO MR. MCDADE

                    MR. MCDADE
But we don't audit for
sickness, we audit for death.

                    DR. KORMAN (O.S.)
Yeah, that's a limitation.

                                   FADE OUT:

EPILOGUE

FADE IN:

PLATE — "In 2010 the Fraser River sockeye had the best return in 35 years."

FADE TO:

PLATE — "But that only meant returns per spawner were back to normal."

FADE TO:

PLATE — "By 2012, sockeye returns were lower than expected."

FADE TO:

PLATE — "By 2016, the sockeye run was one of the worst ever . . ."

FADE TO:

TAIL CREDITS . . .

An electrical glitch, like someone plugging in a device. A voice graph appears for this dialogue.

> MR. TAYLOR (O.S.)
> I object! "Ambush." "Back door." "Derailing." "Breach of rules." "Unfair." "Prejudicial" are all words that apply. The topic that this panel has been
> (MORE)

                    MR. TAYLOR (CONT'D)
          presented for is science and
          the role of science and more
          specifically, the role of
          science and decision-making
          in the department. And most of
          the evidence has stuck to that.
          Mr. McDade seeks to turn this
          part of the Inquiry into an
          investigation on aquaculture,
          which it's not . . .

Another electrical glitch.

                                        FADE OUT:

                        THE END

Acknowledgements

A great big thanks to Commissioner Bruce Cohen, his entire staff, all of the legal counsel, their teams, and all of the participants in the Cohen Inquiry. Special mention to Leonard Giles who served as the Registrar and allowed me to adjust the room and lighting so that I could get the best images possible within the terms of reference. And a particular shout-out to Carla Shore, director of Communications for the Inquiry, for her guidance and help throughout my time at the hearings.

I would like to thank my team for their hard work and dedication during the completion of the film. It was an enormous task to go through all of the testimony and extract the key parts of the Inquiry and distill them into a very watchable film. In particular, I would like to thank Rob Neilson, who pulled together the post-production team and pushed the film over the finish line. This was not an easy film to cut together since there was only one camera allowed in the courtroom and the many sudden camera moves created added challenges for the edit.

A big thanks to Jan Westendorp for her amazing skill and guidance in managing the design and publishing of this book. And to Lesley Cameron, whose terrific and prompt story editing and attention to detail kept the book project moving forward and a dream project to work on. Both of you have incredible attention to detail which continues to amaze me.

I would also like to thank Marcus Bowcott for the wonderful original painting that says so much and became a crucial part of the key art for the film.

I would also like to acknowledge the Government of Canada for embracing public inquiries and the role they play in solving difficult issues that face the nation. So many people rely on healthy wild fish populations for their livelihood or nutrition, and I believe this Inquiry created an opportunity to fix the impacts of open net pen fish farms.

I would also like to acknowledge our Fraser River sockeye populations. You hold a special place in the hearts of British Columbians and I'm sorry we have not done more to protect you. Let's hope that now that we know what the problem is, thanks in part to this Inquiry, we do something about it and allow you to flourish once again.

References

Exhibits

The following list contains the exhibit numbers and descriptions of exhibits that were presented to the Cohen Inquiry.

Exhibit #	Description
395	Fisheries and Oceans Canada. (1988, December). *Fraser River sockeye management and enhancement plan summary prepared by Fraser River Sockeye Task Force for Area Planning Committee.* Government of Canada.
396	Cass, A., Folkes, M., & Pestal, G. (2004). *Methods for assessing harvest rules for Fraser River sockeye salmon.* Canadian Science Advisory Secretariat. Fisheries and Oceans Canada.
397	Cass, A., & Grout, J. (2006). Workshop to assess population dynamics of cyclic Fraser River sockeye and implications for management: February 7–8, 2006, University of British Columbia, Vancouver, BC.
398	Pestal, G., Ryall, P., & Cass, A. (2008.) *Collaborative development of escapement strategies for Fraser River sockeye: Summary report 2003 to 2008.* Department of Fisheries and Oceans Canada.

399   Pestal, G., Huang, A., Cass, A., & the FRSSI Working Group. (2011). *Updated methods for assessing harvest rules for Fraser River sockeye salmon.* Fisheries and Oceans Canada.

400   Staley, M. (2010). *Fraser River Sockeye Spawning Initiative (FRSSI): A review for the Cohen Commission.* [no publisher]

401   Curriculum vitae of Michael Staley.

402   Curriculum vitae of Ken Wilson.

403   Fisheries and Oceans Canada. (2008, April). *A framework for socio-economic analysis to inform integrated fisheries management planning and fish harvest decisions.* Government of Canada.

404   Hague, M., & Patterson, D.A. (2008). *Improvements to environmental management adjustment models, SEF final report.* Fisheries and Oceans Canada, Science Branch and Pacific Salmon Commission.

405   Patterson, D.A., & Hague, M.J. (2008). *Evaluation of potential freshwater factors linked to the decline of Early and Late Stuart sockeye salmon: SEF final report.* Fisheries and Oceans Canada.

407   Fisheries and Oceans Canada. (2010). *Fraser sockeye escapement strategy 2010: Model overview and summary of 2010 planning simulations.* Government of Canada.

409   Pestal, G., Ryall, P., & Cass, A. (2008). *Collaborative development of escapement strategies for Fraser River sockeye: Summary report 2003–2008* [Canadian manuscript report of Fisheries and Aquatic Sciences 2855].

411   Department of Fisheries and Oceans. (2011). *Guidelines for applying updated methods for assessing harvest rules for FRSS.* Government of Canada.

412   Wilson, K., Orr, C., & Young, J. (2007, February 28). Letter to Paul Ryall, re Fraser River Sockeye Spawning Initiative/ WSP pilot.

413   Wilson, K. (2009, March). *Fraser River Integrated Sockeye Spawning Initiative report,* prepared for Upper Fraser Fisheries Conservation Alliance (UFFCA).

414   Curriculum vitae of Dr. James C. Woodley.

415   Curriculum vitae of Dr. Carl Walters.

416   Shinners, C.W., Forrest, M., Schultz, David C., Schmitten, R.A., Smits, T.A., & Manary, E.P. (1985). *Fraser River sockeye forecast.* International Pacific Salmon Fisheries Commission.

417   Walters, C., LeBlond, P., & Riddell, B. (2004). *Does over-escapement cause salmon stock collapse?* Technical Paper. Pacific Fisheries Resource Conservation Council.

418   Gilhousen, P. (1992). *Estimation of Fraser River sockeye escapements from commercial harvest data, 1892–1944.* Volume 27. International Pacific Salmon Fisheries Commission.

419   Clark, R., Willette, M., Fleischman, S., & Eggers, D. (2007). *Biological and fishery-related aspects of overescapement in Alaskan sockeye salmon* Oncorhynchus nerka. Alaska Department of Fish and Game.

420   Fisheries and Oceans Canada. (2010). *Summer run sockeye salmon: Near final escapement estimates report.* Government of Canada.

551   Curriculum vitae of Dr. Scott Hinch.

552   Curriculum vitae of Dr. Eduardo Martins.

553   Hinch, S., & Martins, E. (2011). *A review of potential climate change effects on survival of Fraser River sockeye and an analysis of interannual trends in en route loss and pre-spawn mortality.* Cohen Commission Technical Report 9.

557   Hinch, S.G., & Gardner, J. (Eds.). (2009). *Proceedings of the Conference on Early Migration and Premature Mortality in Fraser River Late-Run Sockeye Salmon.* Vancouver, BC. Pacific Fisheries Resource Conservation Council.

558   Miller, K., Li, S., Kaukinen, K.H., Ginther, N., Hammill, E., Curtis, J.M.R., Patterson, D.A., Sierocinski, T., Donnison, L., Pavlidis, P., Hinch, S.G., Hruska, K.A., Cooke, S.J., English, K.K., & Farrell, A.P. (2011). Genomic signatures predict migration and spawning failure in wild Canadian salmon. *Science, 331*(6014): 214–217.

559   Young, J.L., Hinch, S.G., Cooke, S.J., Crossin, G.T., Patterson, D.A., Farrell, A.P., Van Der Kraak, G., Lotto, A.G., Lister, A., Healey, M.C., & English, K.K. (2006). Physiological and energetic correlates of en route mortality for abnormally early migrating adult sockeye salmon (*Oncorhynchus nerka*) in the Thompson River, British Columbia. *Canadian Journal of Fisheries and Aquatic Sciences, 63*(5): 1067–1077.

560   Walker, I.J., & Sydneysmith, R. (2008). British Columbia. In D.S. Lemmen, F.J. Warren, J. Lacroix, & E. Bush (Eds.), *From impacts to adaptation: Canada in a changing climate 2007* (pp. 329–386). Government of Canada.

561   Farrell, A.P., Hinch, S.G., Cooke, S.J., Patterson, D.A., Crossin, G.T., Lapointe, M., & Mathes, M.T. (2008). Pacific salmon in hot water: Applying aerobic scope models and biotelemetry to predict the success of spawning migrations. *Physiological and Biochemical Zoology, 81*(6): 697–708.

620   Landry, J. (2010, January 5). Memorandum to J. Kristmanson and J. Parsons for the Minister RE: 2009–507–00175 Potential causes of poor returns of Fraser River sockeye salmon: With focus on sea lice impacts.

621   Webb, A. (2010, September 21). Action request from Regional Director of Policy re: Synthesis of evidence from a workshop on the decline of Fraser River sockeye.

622 Davis, T. (2009, October 2). Email to D. Lake, J. Grout, P. Ryall, A. Thomson, A. Bate, M. Saunders, D. Patterson re: Sample speeches and templates.

622A House of Commons. (2009, September 17). Speaking notes for a Member of Parliament for a debate on low returns of sockeye salmon to the Fraser River, Government of Canada.

623 Lake, D. (2009, September 29). Email to multiple parties re: Sample speeches and templates.

624 Webb, A. (2009, October 5). Email string to multiple parties re: Sea lice speech with attached draft speaking notes for a Member of Parliament for a debate on low returns of sockeye salmon to the Fraser River.

625 Webb, A. (2009, October 5). Revised email. T. Robbins, L. Richards re: Sea lice speech with attached draft speaking notes for a Member of Parliament for a debate on low returns of sockeye salmon to the Fraser River.

626 Robbins, T. (2009, October 6). Email to L. Richards re: Revised sea lice speech with Fraser River sockeye overview with attached draft speaking notes for a Member of Parliament for a debate on low returns of sockeye salmon to the Fraser River, House of Commons, September 27, 2009.

627 Miller-Saunders, K. (2009, November 3). Email to L. Brown, M. Saunders, and L. Richards re: A discussion on BC sockeye salmon and science issues.

628 Miller-Saunders, K. (2009, November 4). Email to M. Saunders re: version 2.

633 Davis, T. (2009, September 11). Email to K. Colpitts re: Media lines on Minister's roundtable on Fraser River sockeye. Fisheries and Oceans Canada.

634   Department of Fisheries and Oceans. (2009, September 11).
       Summary of Minister Shea's Pacific Region visit.
       Government of Canada.

635   Fisheries and Oceans Canada. (2008, November 13).
       Briefing note RE: Funding requested for research on links
       between plasmacytoid leukemia and shifts in migration
       timing and high mortality of sockeye salmon in the Fraser
       River.

636   Shea, G. (2011, March 3). Letter to Dr. Alexandra Morton
       RE: Dr. Kristi Miller paper published in *Science*. Minister of
       Fisheries and Oceans.

637   Beamish, D. (Chair). (2009, February 6). *Report on the
       Workshop on Climate Impacts on Pacific Salmon*. Committee
       for Scientific Cooperation. Pacific Salmon Commission.

638   Miller, K. (2010). Proposed research on suspected novel
       virus from genomics study on sockeye salmon [Internal
       document]. Fisheries and Oceans Canada. Pacific
       Biological Station, Nanaimo, BC.

639   Miller-Saunders, K. (2010, April 23). Proposed 2010 DFO
       funded genomics research relating to sockeye declines.
       Fisheries and Oceans Canada. Pacific Biological Station,
       Nanaimo, BC.

640   Thomson, A., Richards, L., Radford, D., & Sprout, P. (2008).
       Memorandum for the Minister. Strategy to address the
       issue of sea lice and salmon farms in Pacific Region.
       Fisheries and Oceans Pacific Region.

641   Davis, T. (2008, January 28). Email to Brian Riddell Re:
       Counteracting the findings of Alexandra Morton's sockeye
       research near the Discovery Islands. Fisheries and Oceans
       Canada, Pacific Region.

642   D'Auray, M. (2008, January 31). Memorandum for the Minister. New research results on the interactions between sea lice and juvenile pink salmon. Fisheries and Oceans Canada, Pacific Region.

643   Davis, T. (2009, August 19). Email to Dr. Laura Richards, A. Thomson, S. Farlinger, and P. Sprout re: Brian Riddell article and media lines. Fisheries and Oceans Canada, Pacific Region.

176   Curriculum vitae of Heather Stahlberg.

PPR10   *Policy and practice report: Overview of Fraser River sockeye net and gross escapement data.* (2011, April 1). Commission of Inquiry into the Decline of Sockeye Salmon in the Fraser River. Government of Canada.

748   Peterman, R.M., & Dorner, B. (2011, February). *Fraser River sockeye production dynamics.* Technical Report 10. Cohen Commission.

749   Curriculum vitae of Randall M. Peterman.

750   Curriculum vitae of Brigitte Dorner.

751   Bodtker, K., Peterman, R., & Bradford, M. (2007, February). Accounting for uncertainty in estimates of escapement goals for Fraser River sockeye salmon based on productivity of nursery lakes in British Columbia, Canada. *North American Journal of Fisheries Management, 27*(1): 286–302.

752   Northwest Fisheries Science Center. (2009, February). *Factors affecting sockeye salmon returns to the Columbia River in 2008.* NOAA Fisheries. Seattle, Washington.

753   DFO Science. (1997, January). *Stock status report, Rivers and Smith Inlet sockeye.* Government of Canada.

754   Peterman, R. (2004). Possible solutions to some challenges facing fisheries scientists and managers. *Journal of Marine Science, 61*(8): 1331–1343.

Y	Walters, C. (n.d.). *Where have all the sockeye gone?* Fisheries Centre, University of British Columbia.
Z	Woodey, J.C. (1996, April 18). Memorandum to L. Loomis and A. Lill, Fraser River Panel re: Assessment of the classification of stocks to stock group.
1284	Curriculum vitae of Dr. Stewart McKinnell.
1285	Curriculum vitae of Dr. Richard Beamish.
1286	Curriculum vitae of Dr. David Welch.
1287	Curriculum vitae of Enrique Curchitser.
1288	Biography of Cornelius Groot.
1289	Curriculum vitae of Masahide Kaeriyama.
1290	Curriculum vitae of Katherine Myers.
1291	McKinnell, S.W., Curchitser, E., Groot, C., Kaeriyama, M., & Myers, K.W. (2011). *The decline of Fraser River sockeye salmon* Oncorhynchus nerka *(Steller, 1743) in relation to marine ecology.* Cohen Commission Technical Report 4.
1292	Welch, D., Melnychuk, M.C., Payne, J.C., Rechisky, E.L., Porter, A.D., Jackson, G.D., Ward, B.R., Vincent, S.P., Wood, C.C., & Semmens, J. (2011). In situ measurement of coastal ocean movements and survival of juvenile Pacific salmon. *PNAS Early Edition.*
1294	Beamish, D., Cummins, P., Hyatt, K., Irvine, J., Masson, D., Miller, K., Neville, C., Pena, A., Sweeting, R., Thomson, R., Trudel, M., Tucker, S., & Whitfield, P. (2011, April). Ocean conditions inside and outside the Strait of Georgia are important contributors to the Fraser sockeye situation, including the high seas. PowerPoint presentation. Fisheries and Oceans Canada, Pacific Region.

1311   Beamish, R.J., Riddell, B.E., Lange, K.L., Farley Jr., E., Kang, S., Nagasawa, T., Radchenko, V., Temnykh, O., & Urawa, S. (2009). *A long-term research and monitoring plan (LRMP) for Pacific salmon* (Oncorhynchus spp.) *in the North Pacific Ocean.* North Pacific Anadromous Fish Commission Special Publication No. 1.

1312   Ware, D.M., & Thomson, R.E. (2005). Bottom-up ecosystem trophic dynamics determine fish production in the northeast Pacific. *Science, 308*(5726): 1280-1284.

1314   Welch, D., Melnychuk, M.C., Rechisky, E.R., Porter, A.D., Jacobs, M.C., Ladouceur, A., McKinley, R.S., & Jackson, G.D. (2009). Freshwater and marine migration and survival of endangered Cultus Lake sockeye salmon smolts using POST, a large-scale acoustic telemetry array. *Canadian Journal of Fisheries and Aquatic Studies, 66*(5): 736–750.

1319   Beamish, R.J., & Riddell, B.E. (2009). The future of fisheries science on Canada's west coast is keeping up with the changes. In R.J. Beamish & B.J. Rothschild (Eds.), *The future of fisheries science in North America. Fish & Fisheries Series,* volume 31 (chapter 29, pp. 567–595). Springer.

1320   Healey, M. (2011). The cumulative impacts of climate change on Fraser River sockeye salmon and implications for management. *Canadian Journal of Fisheries and Aquatic Sciences, 68*(4): 718–737.

1321   Beamish, D., Sweeting, R., Lange, K., & Preikshot, D. (2009). *A possible reason for the poor returns of sockeye salmon to the Fraser River in 2009.* Fisheries and Oceans Canada, Pacific Biological Station, Nanaimo, BC.

1322   Noakes, D.J., Beamish, R.J., Sweeting, R., & King, J. (2000). Changing the balance: Interactions between hatchery and wild Pacific coho salmon in the presence of regime shifts. *North Pacific Anadromous Fish Community Bulletin,* (2): 155–163.

1323   Beamish, R.J., Sweeting, R.M., Lange, K.L., Noakes, D.J., Preikshot, D., & Neville, C.M. (2011). Early marine survival of coho salmon in the Strait of Georgia declines to very low levels. *Marine and Coastal Fisheries, 2*(1): 424–439.

1324   Noakes, D.J., & Beamish, R.J. (2011). Shifting the balance: Towards sustainable salmon populations and fisheries of the future. In W.W. Taylor, A.J. Lynch, & M.G. Schechter (Eds.), *Sustainable fisheries: Multi-level approaches to a global problem* (pp. 23–50). American Fisheries Society.

1325   Chittenden, C.M., Jensen, J.L.A., Ewart, D., Anderson, S., Balfry, S., Downey, E., Eaves, A., Saksida, S., Smith, B., Vincent, S., Welch, D., & McKinley, R.S. (2010). Recent salmon declines: A result of lost feeding opportunities due to bad timing? *PLoS One, 5*(8): e12343.

1326   Crawford, B., & Irvine, J. (2010). *State of the Pacific Ocean 2009.* Fisheries and Oceans Canada.

1327   Crawford, W.R., & Irvine, J.R. (Eds.). (2010). *State of physical, biological, and selected fishery resources of Pacific Canadian marine ecosystems in 2009.* Fisheries and Oceans Canada.

1328   Beamish, D., Neville, C.M., Sweeting, R.M., & Poier, K.L. (2001). *Persistence of the improved productivity of 2000 in the Strait of Georgia, British Columbia, Canada through to 2001.* Fisheries and Oceans Canada, Science Branch—Pacific Region, Pacific Biological Station, Nanaimo, BC.

1448   Curriculum vitae of Michael Kent.

1449   Kent, M. (2011). *Infectious diseases and potential impacts on survival of Fraser River sockeye salmon.* Cohen Commission Technical Report 1.

1451	Curriculum vitae of Stewart Johnson.
1452	Fisheries and Oceans Canada. (2011, May). Organizational Charts of DFO Salmon and Freshwater Ecosystems Division. Pacific Region, Vancouver, BC.
1453	Curriculum vitae, highlights specific to the Cohen Commission mandate, of Craig Stephen.
1454	Stephen, C., Stitt, T., Dawson-Coates, J., & McCarthy, A. (2011). *Assessment of the potential effects of diseases present in salmonid enhancement facilities on Fraser River sockeye salmon.* Cohen Commission Technical Report 1A.
1455	Curriculum vitae of Christine MacWilliams.
1456	Garver, K. (2010). *Hypothesis: Diseases in freshwater and marine systems are an important contributor to the Fraser sockeye situation.* Fisheries and Oceans Canada, Pacific Biological Station, Nanaimo, BC.
1457	Garver, K. (n.d.). Email to Cohen Commission re: IHNV prevalence rates in Fraser River sockeye salmon data and VHSV prevalence in herring and Atlantic salmon. Fisheries and Oceans Canada, Pacific Biological Station, Nanaimo, BC.
1458	MacWilliams, C. (2009). Update on science review. Department of Fisheries and Oceans. Government of Canada.
1459	Specific Pathogen Control Plan for Renibacterium salmoninarum, at B.C. Federal Enhancement Hatcheries and Affiliates [no author, no date].
1460	Williams, C. (2010, September 29). Memorandum to J. Willis et al. re: Broodstock screening results—Lakelse sockeye.
1461	Fisheries and Oceans Canada. (2010). Introduction to pathogens, diseases and host pathogen interactions of sockeye salmon. PowerPoint Presentation. Pacific Biological Station, Nanaimo, BC.

1462   Wagner, G.N., Fast, M.D., & Johnson, S.C. (2007).
        Physiology and immunology of Lepeophtheirus salmonis
        infections of salmonids. *Trends in Parasitology, 24*(4):
        176–183.

1463   Fisheries and Oceans Canada. (2010, December 19).
        Salmonid Enhancement Program Aquaculture Licence.
        Application Form. Government of Canada.

1464   Rolland, J.B., & Winton, J.R. (2003). Relative resistance of
        Pacific salmon to infectious salmon anaemia virus. *Journal
        of Fish Diseases, 26*(9): 511–520.

1465   MacWilliams, C., Johnson, G., Groman, D., & Kibenge, F.S.B.
        (2007). Morphologic description of infectious salmon
        anaemia virus (ISAV)-induced lesions in rainbow trout
        Oncorhynchus mykiss compared to Atlantic salmon *Salmo
        salar*. *Diseases of Aquatic Organisms, 78*(1):1–12.

1466   Freshwater Fisheries Society of BC. (2010, November). Fish
        Health Management Plan Fraser River Trout Hatchery.

1467   Freshwater Fisheries Society of BC. (2008, March). Fish
        Health Management Plan Vancouver Island Trout
        Hatchery.

1468   Freshwater Fisheries Society of BC. (2008, March). Fish
        Health Management Plan Clearwater Trout Hatchery.

1471   Publicly available PCR test results for ISAV in British
        Columbia farmed salmon from 2003–2010. Province of BC.

1472   Jones, S., Kim, E., & Bennett, W. (2008). Early development
        of resistance to the salmon louse, *Lepeophtheirus salmonis*
        (Krøyer), in juvenile pink salmon, *Oncorhynchus gorbuscha*
        (Walbaum). *Journal of Fish Diseases, 31*(8): 591–600.

1473   Jones, S., & Hargreaves, B. (2009). Infection threshold to
        estimate *Lepeophtheirus salmonis*-associated mortality
        among juvenile pink salmon. *Diseases of Aquatic Organisms,
        84*(2): 131–137.

1474   Johnson, S., & Margolis, L. (1993). Efficacy of ivermectin for control of the salmon louse *Lepeophtheirus salmonis* on Atlantic salmon. *Diseases of Aquatic Organisms, 17*(2): 101–105.

1475   Connors, B., Krkosek, M., Ford, J., & Dill, L. (2010). Coho salmon productivity in relation to salmon lice from infected prey and salmon farms. *Journal of Applied Ecology, 47*(6): 1372–1377.

1476   Price, M., Proboszcz, S.L., Routledge, R.D., Gottesfeld, A.S., Orr, C., & Reynolds, J.D. (2011). Sea louse infection of juvenile sockeye salmon in relation to marine salmon farmers on Canada's west coast. *PLoS One, 6*(3): 1–9.

1477   Traxler, G., Roome, J., & Kent, M. (1993). Transmission of infectious hematopoietic necrosis virus in seawater. *Diseases of Aquatic Organisms, 16*: 111–114.

1478   Kent, M.L., Traxler, G.S., Kieser, D., Richard, J., Dawy, S.C., Shaw, R.W., Prosperi-Porta, G., Ketcheson, J., & Evelyn, T.P.T. (1998). Survey of salmonid pathogens in ocean-caught fishes in British Columbia, Canada. *Journal of Aquatic Animal Health, 10*: 211–219.

1480   Callan, T. (2011, May 27). Letter to Gregory McDade. RE: Cohen Commission—(MOE) Response to Gregory McDade's enquiry of FFSBC Fish Health Case 2010–1100.

1481   Prince, M.H.H., Morton, A., & Reynolds, J.D. (2010). Evidence of farm-induced parasite infestations on wild juvenile salmon in multiple regions of coastal British Columbia, Canada. *Canadian Journal of Fisheries and Aquatic Sciences, 67*(12): 1925–1932.

1482   Rimstad, E. (2011). Examples of emerging virus diseases in salmonid aquaculture. *Aquaculture Research, 42*(s1):86–89.

1483   Robertsen, B. (2011). Can we get the upper hand on viral diseases in aquaculture of Atlantic salmon? *Aquaculture Research, 42*(1): 125–131.

1484   Mennerat, A., Nilsen, F., Ebert, D., & Skorping, A. (2010). Intensive farming: Evolutionary implications for parasites and pathogens. *Journal of Evolutionary Biology, 37*(2–3): 59–67.

1485   Reno, P.W. (1988). Factors involved in the dissemination of disease in fish populations. *Journal of Aquatic Animal Health, 10*(2): 160–171.

1486   Walker, P.J., & Winton, J.R. (2010). Emerging viral diseases of fish and shrimp. *Veterinary Research, 41*(6): 51.

1487   Ford, J.S., & Myers, R.A. (2008). A global assessment of salmon aquaculture impacts on wild salmonids. *PLoS Biology, 6*(2): e33.

1488   Kent, M.L., & Dawe, S.C. (1990). Experimental transmission of a plasmacytoid leukemia of chinook salmon, *Oncorhynchus tshawytscha. Journal of Cancer Research, 50*(17 Suppl): 5679S–5681S.

1489   Newbound, G.C., & Kent, M.L. (1991). Experimental interspecies transmission of plasmacytoid leukemia in salmonid fishes. *Diseases of Aquatic Organisms, 10*: 159–166.

1490   Eaton, W.D., & Kent, K. (1992). A retrovirus in chinook salmon (*Oncorhynchus tshawytscha*) with plasmacytoid leukemia and evidence for the etiology of the disease. *Journal of Cancer Research, 52*(23): 6496–6500.

1491   Stephen, C., Ribble, C.S., & Kent, M.L. (1996). Descriptive epidemiology of marine anemia in seapen-reared salmon in southern British Columbia. *Canadian Veterinary Journal, 37*(7): 420–425.

1492   Stephen, C., & Ribble, C. (1995). The effects of changing demographics on the distribution of marine anemia in farmed salmon in British Columbia. *Canadian Veterinary Journal, 36*(9): 557–562.

1493   Eaton, W.D., Folkins, B., & Kent, M.L. (1994). Biochemical and histologic evidence of plasmacytoid leukemia and salmon leukemia virus SLV in wild-caught chinook salmon *Oncorhynchus tshawytscha* from British Columbia expressing plasmacytoid leukemia. *Diseases of Aquatic Organisms, 19*(2): 147–151.

1494   Kent, M. (n.d.). The impacts of diseases on pen-reared salmonids on coastal marine environments. *Fisken og Havet, 13.* In press.

1495   St-Hilaire, S., Ribble, C.S., Stephen, C., Anderson, E., Kurath, G., & Kent, M.L. (2002). Epidemiological investigation of infectious hematopoietic necrosis virus in salt water net-pen reared Atlantic salmon in British Columbia, Canada. *Aquaculture, 212*(1–4): 49–67.

1496   Saksida, S. (2006). Infectious haematopoietic necrosis epidemic (2001 to 2003) in farmed Atlantic salmon *Salmo salar* in British Columbia. *Diseases of Aquatic Organisms, 72*(3): 213–223.

1497   Province of British Columbia. (2010). BC Salmon Farming Database [Excel spreadsheet].

1498   Province of British Columbia. (2008). BC Salmon Farming Database [Excel spreadsheet].

1499   Fisheries and Ocean Canada. (2010). Program for Aquaculture Regulatory Research (PARR) Calls for Proposals (2010/11), PAAR Project Proposal 2010/11. DFO, Pacific Region.

1500   Miller-Saunders, K. (2011, July 29). Email to Christine MacWilliams. Subject: testing of Atlantic salmon.

1501   Miller-Saunders, K. (2011, July 29). Email to Stewart Johnson. Subject: testing Atlantic salmon for parvovirus.

1502   Vike, S., Nylund, S., & Nylund, A. (2009). ISA virus in Chile: Evidence of vertical transmission. *Archives of Virology, 154*(1): 1–8.

1503	Province of British Columbia. (2002). BC Salmon Farmer Database [Excel spreadsheet].
1504	Province of British Columbia. (2003). BC Salmon Farmer Database [Excel spreadsheet].
1505	Province of British Columbia. (2004). BC Salmon Farmer Database [Excel spreadsheet].
1506	Province of British Columbia. (2005). BC Salmon Farmer Database [Excel spreadsheet].
1507	Province of British Columbia. (2006). BC Salmon Farmer Database [Excel spreadsheet].
1508	Province of British Columbia. (2007). BC Salmon Farmer Database [Excel spreadsheet].
1509	Province of British Columbia. (2009). BC Salmon Farmer Database [Excel spreadsheet].
1510	Curriculum vitae of Dr. Kristi Miller.
1511	Curriculum vitae of Dr. Kyle Garver.
1512	Fisheries and Oceans Canada. (2010, June). Hypothesis: Genomic data indicate that a potentially novel disease, possibly viral in origin, has been affecting a high proportion of juvenile and adult Fraser River sockeye salmon that may weaken fish and directly or indirectly enhance mortality of both smolts and adults. Hypothesis prepared for Pacific Salmon Commission meeting.
1513	Miller, K. (2011, April 15). *Genomic studies suggest that a novel disease is affecting sockeye and may be an important contributor to the Fraser River sockeye situation* [Workshop presentation]. DFO Workshop on Fraser Sockeye Salmon. Fisheries and Ocean Canada.
1514	Chen, E.C., Miller, S.A., DeRisi, J.L., & Chiu, C.Y. (2011). Using a pan-viral microarray assay (Virochip) to screen clinical samples for viral pathogens. *Journal of Visualized Experiments*, (50): 2536.

1515   Garver, K. (2009, October 8). Email to Kristi Miller-Saunders re: Minister memo draft. Fisheries and Oceans Canada, Pacific Biological Station, Nanaimo, BC.

1516   Memorandum for the Minister. (n.d.). Epidemic of a novel, cancer-causing viral disease may be associated with wild salmon declines in B.C. [Internal document]. Fisheries and Oceans Canada.

1517   Miller, K. (2011, May 19). *Timeline of genomic research relating to the mortality-related genomic signature hypothesized to be associated with a potentially novel virus.* Fisheries and Oceans Canada, Pacific Biological Station, Nanaimo, BC.

1518   Garver, K. (2011). *Hypothesis: Diseases in freshwater and marine systems are an important contributor to the Fraser sockeye situation.* Fisheries and Oceans Canada. Pacific Biological Station, Nanaimo, BC.

1519   Garver, K., Tang, P., Hawley, L., & Richard, J. (n.d.). *Microarray-base detection of fish viruses.* Fisheries and Oceans Canada and BC Centre for Disease Control.

1520   Miller-Saunders, K. (2011, April). *2007 versus 2008 genomics contrast study.* Fisheries and Oceans Canada. Pacific Biological Station, Nanaimo, BC.

1521   Miller-Saunders, K. (2010, June). *Hypothesis: Genomic studies suggest that some disease has infected sockeye and has become an important contributor to the Fraser River sockeye situation.* Fisheries and Oceans Canada. Pacific Biological Station, Nanaimo, BC.

1522   Miller-Saunders, K. (2011, May 19). *Timeline of genomic research relating to the mortality-related genomic signature hypothesized to be associated with a potentially novel virus.* Fisheries and Oceans Canada. Pacific Biological Station, Nanaimo, BC.

1523   Miller-Saunders, K. (2009). Epidemic of a novel, cancer-causing viral disease may be associated with wild salmon declines in BC [Unpublished paper].

1524   Miller-Saunders, K. (2009, September 27). *Epidemic of a novel, cancer-causing viral disease may be associated with wild salmon declines in B.C.* Fisheries and Oceans Canada. Pacific Biological Station, Nanaimo, BC.

1525   Miller-Saunders, K. (2009, October 5). Email to Mark Saunders RE: Briefing Report. Fisheries and Oceans Canada. Pacific Biological Station, Nanaimo, BC.

1526   Johnson, S. (2011, March 14). Email to Kyle Garver, Simon Jones RE: Cohn [*sic*] Commission Information for Laura Richards. Fisheries and Oceans Canada.

1527   Kaukinen, K. (2010). Aquaculture Collaborative Research and Development Program (ACRDP) Application Form re Creative Salmon Ltd. Bi-02 in the Molecular Genetics Laboratory, Pacific Biological Station. Fisheries and Oceans Canada. Nanaimo, BC.

1528   Miller-Saunders, K. (2008, November 12). Email to Kyle Garver re: briefing note, with attachment. Head, Molecular Genetics, Pacific Biological Station. Fisheries and Oceans Canada. Nanaimo, BC.

1529   Garver, K., Grant, A., Richard, J., Stucchi, D., & Foreman, M. (2011). *Risks of infectious hematopoietic necrosis virus (IHNv) dispersion associated with Atlantic salmon net pen aquacultures.* Pacific Biological Station, Nanaimo, BC, and Institute of Ocean Sciences, Sidney, BC. Fisheries and Oceans Canada.

1530   Garver, K. (2010, April). *Discovery Islands Modeling Progress Report. RE: Viral shedding.* Fisheries and Oceans Canada. Pacific Biological Station, Nanaimo, BC.

1531   Miller, K. (2010, November). Memorandum for the Minister. Indications of a possibility of infectious diseases associated with poor survival of southern BC salmon stocks. Fisheries and Oceans Canada. Pacific Biological Station, Nanaimo, BC.

1532   Miller-Saunders, K. (2011, June 27). Email to Gary Marty RE: Final "unblinded" FR sockeye histopathology results, 2011–2111. Fisheries and Oceans Canada, Pacific Biological Station, Nanaimo, BC.

1533   Richards, L. (2011, January 11). Email to Kristi Miller-Saunders re: Media requests—science paper. Regional Director Science. Fisheries and Oceans Canada, Pacific Biological Station, Nanaimo, BC.

1534   Curriculum vitae of Josh Korman.

1535   Curriculum vitae of Don Noakes.

1536   Noakes, D.J. (2011). *Impacts of salmon farms on Fraser River sockeye salmon: Results of the Noakes investigation.* Cohen Commission Technical Report 5C.

1538   Noakes, D. (2011, August 10). Memorandum: Re Noakes response to Connors Cohen Commission Technical Report 5B. Thompson Rivers University, Department of Mathematics and Statistics. Kamloops, BC.

1539   Curriculum vitae of Larry Dill.

1540   Dill, L. (2011). *Impacts of salmon farms on Fraser River sockeye salmon: Results of the Dill investigation.* Cohen Commission Technical Report 5D.

1541   Curriculum vitae of Brendan Connors.

1542   Connors, B. (2011, July 27). Memorandum: Re: Response to Noakes' criticisms of Connors' statistical analysis in Cohen Commission Technical Report 5B. School of Resource and Environmental Management, Simon Fraser University, Burnaby, BC.

PPR 20   *Policy and practice report: Aquaculture regulations in B.C.* (2011, July 28). Government of Canada.

1543   Korman, J. (2011). *Summary of information for evaluating impacts of salmon farms on survival of Fraser River sockeye salmon.* Cohen Commission Technical Report 5A.

1544   Korman, J. (2011, May). Excel workbook re Salmon farm fish mortality data. Ecometric Research Inc, Vancouver, BC.

1545   Connors, B. (2011). *Examination of relationships between salmon aquaculture and sockeye salmon population dynamics.* Cohen Commission Technical Report 5B.

1547   Province of British Columbia. (2009). Excel workbook. RE: Atlantic salmon stomach contents, Atlantic salmon escapes.

1548   Province of British Columbia. (2011). List of B.C. Salmon Farmers databases provided to Korman. Cohen Commission, Vancouver, BC.

1549   Province of British Columbia. (2004, July 31). Re: Histopathology report of Slide 61, A3.1-27-1 [Excel spreadsheet].

1550   Fisheries and Oceans Canada. Atlantic salmon escape data [Excel spreadsheet].

1551   Hoenig, J.M., & Heisey, D.M. (2001). The abuse of power: The pervasive fallacy of power calculations for data analysis. *American Statistician*, 55(1): 1–6.

1552   Table showing correlation between BC population and farm fish production as a critique of the Connors Technical report 5B. Cohen Commission.

1553   Morton, A., Routledge, R., & Krkosek, M. (2008). Sea louse infestation in wild juvenile salmon and Pacific herring associated with fish farms off the east-central coast of Vancouver Island, British Columbia. *North American Journal of Fisheries Management*, 28(2):523–532.

1554   Connors, B.M., Krkosek, M., Ford, J., & Dill, L.M. (2010). Coho salmon productivity in relation to salmon lice from infected prey and salmon farms. *Journal of Applied Ecology, 47*(6): 1372–1377.

1555   Marty, G.D., Saksida, S.M., & Quinn, T.J. (2010, December 13). Relationship of farm salmon, sea lice, and wild salmon populations. *PNAS, 107*(52): 22599–22604.

1556   Krkosek, M., Connors, B.M., Morton, A., Lewis, M.A., Dill, L.M., & Hilborn, R. (2011). Effects of parasites from salmon farms on productivity of wild salmon. *PNAS Early Edition*, 1–5.

1557   Morton, A., Routledge, R., McConnell, A., & Krkosek, M. (2011). Sea lice dispersion and salmon survival in relation to salmon farm activity in the Broughton Archipelago. *ICES Journal of Marine Science, 68*(1): 144–156.

1558   Johnson, S.C., & Albright, L.J. (1992). Comparative susceptibility and histopathology of the response of naive Atlantic, chinook and coho salmon to experimental infection with *Lepeophtheirus salmonis* (Copepoda: Caligidae). *Diseases of Aquatic Organisms, 14*:179-193.

1559   Johnson, S.C., Blaylock, R.B., Elphick, J., & Hyatt, K.D. (1996). Disease induced by the sea louse (*Lepeophtheirus salmonis*) (Copepoda: Caligidae) in wild sockeye salmon (*Oncorhynchus nerkai*) stocks of Alberni Inlet, British Columbia. *Canadian Journal of Fisheries and Aquatic Sciences, 53*(12): 2888–2897.

WW   Beamish, D. (2011). *Assessing the impact of salmon farming on Pacific salmon at population level in British Columbia* [Unpublished].

1560   BC Ministry of Agriculture and Lands. (2009). *Annual report: Fish health program.* Province of BC.

1561 Hammel, L., Stephen, C., Bricknell, I., Evensen, Ø., & Bustos, P. (2009). *Salmon Aquaculture Dialogue Working Group report on salmon disease.* Commissioned by the Salmon Aquaculture Dialogue.

1562 Province of British Columbia. (2010). Fish farm population data excel spreadsheet 2000–2010.

1563 Living Oceans. (2009, June). Salmon farm migration map: BC map of salmon farms and wild salmon migration routes.

1564 Supplemental appendices to the annual report Fish Health program.

1565 Province of British Columbia. (2009). Excel workbook re: Fish health on BC salmon farms 2008-2009.

1566 Fisheries and Oceans Canada. (1984) (revised 2004). Fish Health Protection Regulations: Manual of Compliance. *Fish. Mar. Serv. Misc. Spec. Publ.* 31 (Revised).

1567 Miller, O., & Cipriano, R.C. (2003, April). International response to infectious salmon anemia: Prevention, control and eradication. Technical Bulletin No. 1902. USDA-Animal and Plant Health Inspection Service's Veterinary Services. Riverdale, MD, USA.

1568 Marty, G. (2011, June 27). Email to Kristi Miller-Saunders re: Final unblinded FR sockeye histopathology results 2011–2111.

1569 Noakes, D. (1996, December 11). Email to J. Davis, I. Price, and R. Ginetz. RE: DFO Fish Health Submissions to EA Review.

1570 Connors, B.M., Hargreaves, N.B., Jones, S.R.M., & Dill, L.M. (2010). Predation intensifies parasite exposure in a salmonid food chain. *Journal of Applied Ecology*, 47(6): 1365–1371.

1571   Costello, M. (2009). How sea lice from salmon farms may cause wild salmonid declines in Europe and North America and be a threat to fishes elsewhere. *Proceedings of the Royal Society, 276*(1672): 3385–3394.

1572   Barker, D.E., Braden, L.M., Coombs, M.P., & Boyce, B. (2009). Preliminary studies on the isolation of bacteria from sea lice, *Lepeophtheirus salmonis*, infecting farmed salmon in British Columbia, Canada. *Parasitology Research, 105*(4): 1173–1177.

1573   Nylund, A., Hovland, T., Hodneland, K., Nislen, F., & Lovik, P. (1994). Mechanisms for transmission of infectious salmon anaemia (ISA). *Diseases of Aquatic Organisms, 19*:95–100.

1574   Penston, M.J., McBeath, A.J.A., & Miller, C.P. (2011). Densities of planktonic *Lepeophtheirus salmonis* before and after an Atlantic salmon farm relocation. *Aquaculture Environment Interactions, 1*: 225–232.

1576   Fisheries and Oceans Canada. (2011). Draft of Pacific Aquaculture Regulations—Approach on the use of light. Government of Canada.

1759   Curriculum vitae of Simon Richard Macrae Jones.

1760   Curriculum vitae of Dr. Craig Orr.

1761   Curriculum vitae of Michael H.H. Price.

1762   Redacted curriculum vitae of Sonja Saksida, BSC, DVM, MSC.

1763   Yazawa, R., Yasuikle, M., Leong, J., von Schalburg, K.R., Cooper, G.A., Beetz-Sargent, M., Robb, A., Davidson, W.S., Jones, S.R.M., & Koop, B.F. (2008). EST and mitochondrial DNA sequences support a distinct Pacific form of salmon louse *Lepeophtheirus salmonis*. *Marine Biotechnology, 10*(6): 741–749.

1764   Jones, S. (2010). *Aquaculture response: Sea lice, either naturally occurring or passed from fish farms, are an important contributor to the Fraser sockeye situation.* Fisheries and Oceans Canada, Pacific Biological Station, Nanaimo, BC.

1765   Jones, S., & Prosperi-Porta, G. (2011). The diversity of sea lice (Copepoda: Caligidae) parasitic on threespine stickleback (*Gasterosteus aculeatus*) in coastal British Columbia. *Journal of Parasitology, 97*(3): 399–405.

1766   Jones, S., Prosperi-Porta, G., Kim, E., Callow, P., & Hargreaves, N.B. (2006). The occurrence of *Lepeophtheirus salmonis* and *Caligus clemensi* (Copepoda: Caligidae) on three-spine stickleback *Gasterosteus aculeatus* in coastal British Columbia. *Journal of Parasitology, 92*(3):473–480.

1767   Jones, S. Kim, E., & Dawe, S. (2011). Experimental infections with *Lepeophtheirus salmonis* (Krøyer) on threespine sticklebacks, *Gasterosteus aculeatus* L. and juvenile Pacific salmon, *Oncorhynchus spp. Journal of Fish Diseases, 29*(8): 489–495.

1768   Sutherland, B.J.G., Jantzen, S.G., Sanderon, D.S., Koop, B.F., & Jones, S.R.M. (2011). Differentiating size-dependent responses of juvenile pink salmon (*Oncorhynchus gorbuscha*) to sea lice (*Lepeophtheirus salmonis*) infections. *Comparative Biochemistry and Physiology Part D Genomics Proteomics, 6*(2): 213–223.

1769   Beamish, R., Gordon, E., Wasde, J., Pennell, B., Neville, C., Lange, K., Sweeting, R., & Jones, S. (2011). The winter infection of sea lice on salmon in farms in a coastal inlet in British Columbia and possible causes. *Journal of Aquaculture Research and Development, 2*(1): 107.

1770   Jones, R.M.S. (2009). Controlling salmon lice on farmed salmon and implications for wild salmon. *CAB Reviews: Perspectives in Agriculture, Veterinary Science, Nutrition and Natural Resources, 4*(048): 1–13.

1771   Beamish, R., Wade, J., Pennell, W., Gordon, E., Jones, S., Neville, C., Lange, K., & Sweeting, R. (2009). A large, natural infection of sea lice on juvenile Pacific salmon in the Gulf Islands area of British Columbia, Canada. *Aquaculture, 297*(1–4): 31–37.

1772   Brooks, K.M., & Jones, S.R.M. (2008). Perspectives on pink salmon and sea lice: Scientific evidence fails to support the extinction hypothesis. *Reviews in Fisheries Science, 16*(4): 403–412.

1773   Jones, S.R.M., & Hargreaves, N.B. (2007). The abundance and distribution of *Lepeophtheirus salmonis* (Copepoda: Caligidae) on pink (*Oncorhynchus gorbuscha*) and chum (*O. keta*) salmon in coastal British Columbia. *Journal of Parasitology, 93*(6): 1324–1331.

1774   Jones, S.R.M., Wosniok, W., & Hargreaves, N.B. (2006). The salmon louse *Lepeophtheirus salmonis* on salmonid and non-salmonid fishes in British Columbia. *Proceedings of the 11th International Symposium on Veterinary Epidemiology and Economics.*

1775   Beamish, R.J., Jones, S., Neville, C., Sweeting, R., Karreman, G., Saksida, S., & Gordon, E. (2006). Exceptional marine survival of pink salmon that entered the marine environment in 2003 suggests that farmed Atlantic salmon and Pacific salmon can coexist successfully in a marine ecosystem on the Pacific coast of Canada. *ICES Journal of Marine Science, 63*(7): 1326–1337.

1776   Jones, S., & Nemec, A. (2004). *Pink salmon action plan: Sea lice on juvenile salmon and on some non-salmonid species in the Broughton Archipelago in 2003.* Fisheries and Oceans Canada, Pacific Biological Station, Nanaimo, BC.

1777   Johnson, S. (2010). Program for Aquaculture Regulatory Research (PARR) Call for Proposals (2010/2011) Project Proposal 2010/11. Fisheries and Oceans Canada, Pacific Biological Station, Nanaimo, BC.

1778   Brooks, K.M. (2005). The effects of water temperature, salinity and currents on the survival and distribution of the infective copepodid stage of sea lice (*Lepeophtheirus salmonis*) originating on Atlantic salmon farms in the Broughton Archipelago of British Columbia, Canada. *Reviews in Fisheries Science, 13*(3):177–204.

1779   Brooks, K.M., & Stucchi, D.J. (2006). The effects of water temperature, salinity and currents on the survival and distribution of the infective copepodid stage of the salmon louse (*Lepeophtheirus salmonis*) originating on Atlantic salmon farms in the Broughton Archipelago of British Columbia, Canada (Brooks, 2005)—A response to the rebuttal of Krkosek et al. (2005a). *Reviews in Fisheries Science, 14*(1-2): 13–23.

1780   Jones, S. (2010). *Hypothesis: Sea lice, either naturally occurring or passed from fish farms, are an important contributor to the Fraser sockeye situation.* Fisheries and Oceans Canada, Pacific Biological Station, Nanaimo, BC.

1781   Saksida, S. (2010, April 23). Email to Mark Saunders, Laura Brown, Mark Sheppard, Ian Keith, Gary Marty, Brent Hargreaves, Simon Jones, Richard Beamish, and Andrew Thomson RE: Sockeye salmon health program and Marine Harvest interest in sea lice study. Centre for Aquatic Health Sciences.

1782   Saksida, S., & Downey, E. (2008, January 31). Memorandum to M. Thorsrud, Cermaq ASA Re: Overview of sea lice issues and risks for farmed and wild salmon in British Columbia. BC Centre for Aquatic Health.

1783   Parker, P. (2006, October 16). Letter to Dr. Sonja Saksida re: BC Pacific Salmon Forum's Science Advisory Committee funding not approved. BC Pacific Salmon Forum, Nanaimo, BC.

1784   Johnson, S. (2011, March 28). Email to Dr. Sonja Saksida Re: Rebuttal for Price Paper. Head, Aquatic Animal Health, Pacific Biological Station, Fisheries and Oceans Canada, Nanaimo, BC.

1785   Coastal Alliance for Aquaculture Reform and Marine Harvest Canada. (2009, November 18). *Proceedings of the Morbidity/Mortality Effects of Sea Lice on Juvenile Salmon Workshop*. Simon Fraser University, Burnaby, BC.

1786   Orr, C. (2007). Estimated sea louse egg production from Marine Harvest Canada farmed Atlantic salmon in the Broughton Archipelago, British Columbia 2003–2004. *North American Journal of Fisheries Management, 27*: 187–197.

1787   Krkosek, M., Bateman, A., Proboszcz, S., & Orr, C. (2010). Dynamics of outbreak and control of salmon lice on two salmon farms in the Broughton Archipelago, British Columbia. *Aquaculture Environment Interactions, 1(2)*: 137–146 [proof].

1788   B.C. Centre for Aquatic Health. (2010, October). *Sea lice presence and pathogenicity in the Campbell River and Sunshine Coast salmon farming regions of British Columbia*.

1789   B.C. Centre for Aquatic Health Sciences. (2011, March). Sea lice presence and farm production on 120 farms in British Columbia. Addendum to "Sea lice presence and pathogenicity in the Campbell River and Sunshine Coast farming regions of British Columbia."

1790 Beamish, R.J., Jones, S., Neville, C., Sweeting, R., Karreman, G., Saksida, S., & Gordon, E. (2006). Exceptional marine survival of pink salmon that entered the marine environment in 2003 suggests that farmed Atlantic salmon and Pacific salmon can coexist successfully in a marine ecosystem on the Pacific coast of Canada. *ICES Journal of Marine Science.* In press copy.

1791 Downey, E., Eaves, A., Saksida, S., Anderson, S., & Ewart. D. (2009, December). *Discovery Passage plankton monitoring and juvenile salmon assessment 2009.* Report for Campbell River Salmon Foundation, Pacific Salmon Foundation, Marine Harvest Canada, Grieg Seafood, and Mainstream Canada.

1792 Saksida, S., Constantine, J., Karreman, G.A., & Donald, A. (2007). Evaluation of sea lice abundance levels on farmed Atlantic salmon located in the Broughton Archipelago of British Columbia from 2003 to 2005. *Aquaculture Research,* 38(3): 219–231.

1793 Saksida, S., Constantine, J., Karreman, G.A., Neville, C., Sweeting, R., & Beamish, R. (2006). Evaluation of sea lice, *Lepeophtheirus salmonis,* abundance levels on farmed salmon in British Columbia, Canada. *Proceedings of the 11th International Symposium on Veterinary Epidemiology and Economics.*

1794 Saksida, S., Morrison, D., & Revie, C.S. (2010). The efficacy of emamectin benzoate against infestation of sea lice, *Lepeophtheirus salmonis,* on farmed Atlantic salmon, *Salmo salar* L. in British Columbia. *Journal of Fish Diseases,* 33(11): 913–917.

1795 Nendick, L., Sackville, M., Tang, S., Brauner, C.J., & Farrell, A.P. (2011). Sea lice infection of juvenile pink salmon (*Oncorhynchus gorbuscha*): Effects on swimming performance and postexercise ion balance. *Canadian Journal of Fisheries and Aquatic Sciences, 68*: 241–249.

1796 Saksida, S., Thorburn, M.A., Speare, D.J., Markham R.J.F., & Kent, M. (1999). A field evaluation of an indirect immunofluorescent antibody test developed to diagnose plasmacytoid leukemia in chinook salmon (*Oncorhynchus tshawytscha*). *Canadian Journal of Veterinary Research, 63*(2): 107–112.

1797 McDaniels, T.L., Keen, P.L., & Dowlatabadi, H. (2006). Expert judgments regarding risks associated with salmon aquaculture practices in British Columbia. *Journal of Risk Research, 9*(7): 775–800.

Other References

Galloway, G. (2014, January 7). Purge of Canada's fisheries libraries a "historic" loss, scientists say. *Globe and Mail.*

Hammell, L., Stephen, C., Bricknell, I., Evensen, Ø., & Bustos, P. (2009). *Salmon Aquaculture Dialogue Working Group report on salmon disease.* Report commissioned by the Salmon Aquaculture Dialogue.

Linnitt, C. (2013). Harper's attack on science: No science, no evidence, no truth, no democracy. *Academic Matters: OCUFA's Journal of Higher Education.*

www.ingramcontent.com/pod-product-compliance
Lightning Source LLC
Chambersburg PA
CBHW052110030426
42335CB00025B/2915